Ground Truth

Ground
TRUTH

A Guide to Tracking Climate Change at Home

MARK L. HINELINE

With Illustrations by the Author

The University of Chicago Press

Chicago and London

The University of Chicago Press, Chicago 60637
The University of Chicago Press, Ltd., London
© 2018 by The University of Chicago
Published 2018
Printed in the United States of America

27 26 25 24 23 22 21 20 19 18   1 2 3 4 5

ISBN-13: 978-0-226-34794-3 (cloth)
ISBN-13: 978-0-226-34813-1 (paper)
ISBN-13: 978-0-226-34827-8 (e-book)
DOI: https://doi.org/10.7208/
chicago/9780226348278.001.0001

Library of Congress Cataloging-in-Publication Data
Names: Hineline, Mark L., author.
Title: Ground truth : a guide to tracking climate change
at home / Mark L. Hineline, author, illustrator.
Description: Chicago ; London : The University of
Chicago Press, 2018. |
Includes index.
Identifiers: LCCN 2017046037 | ISBN 9780226347943
(cloth : alk. paper) | ISBN 9780226348131 (pbk. : alk.
paper) | ISBN 9780226348278 (e-book)
Subjects: LCSH: Climatic changes.
Classification: LCC QC903 .H55 2018 | DDC 577.2/2—dc23
LC record available at https://lccn.loc.gov/2017046037

♾ This paper meets the requirements of ANSI/NISO
Z39.48-1992 (Permanence of Paper).

For my granddaughter,
Evee

Contents

Preface

As did many Americans, I first grew alarmed about global warming in 1989, when Bill McKibben published *The End of Nature*. I already knew that warming, or climate change, was a problem. But McKibben, along with the conversations that his book spawned, provided a framework for thinking more deeply about this emerging environmental threat. My concerns deepened when, in 1995, I began to teach a course titled History of Environmentalism at the University of California, San Diego, with readings from Henry David Thoreau and Aldo Leopold. It was while preparing the lectures for that course that I first encountered the strange, not-yet-quite-really-a-science, *phenology*, the observation and study of biological cycles having cues in climates and seasons. Within a year, I began to assign the writing of a phenological journal as part of the course. Students chose a patch of nature, identified the plants growing there and any animals that passed through, described them (and any changes, although there is sometimes little change during a Southern California academic term), and at the same time reflected on course readings and discussions.

My interest in phenology and in seasonality broadened when I began to explore a landscape new to me, the Sonoran Desert, where the plants and animals seemed to wear their phenologies on their sleeves and seasons have peculiarities I'd not before encountered, such as the "monsoonal" rains, thunderstorms, and dust storms of Sonoran summer. Guided by the books of Gary Paul Nabhan, Ann Zwinger, and Ed Abbey, I grew to love the region and to enjoy its intersecting phenophases and phenological cues.

Since that time, phenology has shed some of its obscurity, almost entirely due to its relevance for tracking and confirming the consequences of anthropogenic climate change for Earth's biota. Today, a host of websites is devoted to aspects of phenology. Almost all of them begin with an introductory paragraph designed to answer the question "what is phenology?" If the present book does nothing else, I hope that it answers that question with sufficient clarity and verve that "phenology" becomes as common in the vocabulary of environmentalists and amateur naturalists as "paleontology," "hydrology," or any of dozens of terms that designate areas of scientific observation, inquiry, and expertise.

As a way of opening new doors to an understanding of climate change, including changes that are unfolding even as you read this, *Ground Truth* turns to phenology and to related areas, such as seasonality (which deals with physical cycles over the course of a year, as opposed to phenology, which is about annual biological cycles). Both of these are our present reality. This book is not designed to convince you or anyone else that climate change is occurring or that it is caused by human activity. Of course it is occurring. Of course it is caused by human activity. There are presently hundreds of books devoted entirely to showing this. I won't add to that glut. I might as well exhaust our energies proving that day comes after night.

This book is not a textbook. Neither is it a comprehensive manual for pursuing phenology as a citizen scientist, although I fervently hope that citizen phenologists find in these pages a welcome companion. Rather, it is about three things: climate change, seasons, and phenology. I hope that anyone with interests in birds, rocks, trees, wind, butterflies, deer, beetles, blue skies, flowers, and the rest of the natural world will find something of interest here.

Part 1 of the book introduces phenology as a way of coming to terms with climate change; it prepares you to make phenological observations and keep records. In the first chapter, you will learn about the value of phenological observations for paying attention to your own place in history—and in nature. Chapter 2 shows that anthropocentric—human-caused—climate change is a developing phenomenon, and you will learn about connections between seasons and climate change. The third chapter is a brief history of phenology. In the fourth, I talk about observing, record keeping, and the uses to which phenological observations can be put. The fifth chapter examines the nature of change itself and provides some quick guidelines as to how you might measure the changes in your local landscape that will unfold in years to come.

The five chapters that make up part 2 look to a set of essential details for following the phenologies of plants, invertebrates and amphibians, birds, and mammals. There is also a chapter on weather, which is not, strictly speaking, a subject of phenological study but is closely related to it.

These chapters are no more than introductions to their several subjects. Those who already watch birds, know their plants, or keep a weather station will find them rather too elementary. My goal is to whet appetites and to offer some guidance, not to provide a definitive overview. Neither have I made an attempt to organize each of these chapters following a unified plan, wherein the contents of one chapter correspond to the next. Instead, I have let the subject matter shape the organization of each chapter.

I discuss the ranges of plants and animals, sometimes by naming states where they are present; but in cases where they range over many states, I will mention the states from which they are absent or mostly absent. I also sometimes mention regions, such as the northern Great Plains or the Mississippi River Valley. More precise range maps can be found in guides to wildlife and plants, and these are listed in the bibliographic essay. It makes no sense to duplicate these here.

In discussions that contain phenological dating, for instance, "late fall to early winter," or "November to early February," the range of dates often reflects the extent of the geographical range, where "late fall" events tend to occur in the north and "early winter" events farther south. In the same way, "late winter to early spring" events range from south to north.

Part 3 consists of a single chapter, in which I raise larger issues with respect to anthropogenic climate change.

I present two different conceptions about phenological observation in this book and have taken pains to keep them separate without opposing either of them to the other. For centuries, even before there was

a word to describe what they were doing, people from all stations of
life have tracked seasonal cycles in plants and animals and have made
entries in notebooks about what they've observed. I discuss some of
them in chapter 3. For the most part, these were amateurs, with an
interest in describing local phenologies, a curiosity entirely in keeping
with their zeal for discovering local plants and animals and even fossils;
describing them; thinking of their home as part of a natural tapestry
or a symphony. I have attempted to prepare the groundwork for their
present-day successors, the people who wish to catalog and to know
what I call (after the practice in parts of New England) their dooryards.
When I speak of your *dooryard*, by the way, I'm making a technical point
about the difference between that place and the *commons*, another New
England land-use designator. The commons is shared, whereas your
dooryard is in some ways yours and yours alone, in your thoughts if not
in property law.

In addition to individual interest, there is another reason to keep records of phenological events. The appearance of flowers and the "hatching" of insects provide valuable data by which scientists can track climatic changes as well as build predictive models for understanding
how nature is currently changing. These scientists are very much in the
business of developing models with which they can understand and
predict ecological change. In more than one place, I put a diminished
emphasis on using phenological observations and analysis as a way of
making predictions, whether of future climatic changes or more specifically about ecological or environmental changes. This is not because
I do not recognize the value of predictions, or that of the models being
developed in order to make predictions. These will grow more sophisticated over time and will play important roles in mitigating environmental damage from climate change, when this is possible—damage
that will include the extinctions of species. My lack of emphasis here
regarding the predictive value of phenological observations reflects a
choice of focus: I want to highlight observation while bracketing the
purposes to which observations might be put—the knowledge on which
they may bear. This is more than what scientists once called Baconian-
ism and still sometimes call "mere description." But it is, in fact, at least

that. Accurate observations, carefully recorded and (in some cases) reported, have a beauty *and* a rigor unto themselves. More than that, the overall drift in this book is toward making up one's own mind about the relationship between anthropogenic climate change and the nature through which one passes every day.

It may bemuse some that this book was written by a historian of science rather than by a working scientist. In pursuing these topics—climate change and phenology—I follow on a trail broken earlier by several of my colleagues, who have seized on climate change, the sciences, and the social institutions that surround a concern for the earth and its environment, with an energy and deep concern that I hope also inspires others in years to come, as they have inspired me.

PART | 1

Prologue

Before going past this page of the book, we need to define a term. So, may I ask for a quick indulgence? Please, step outside for a walk. I don't wish to be imperious, certainly not on our first meeting, so I won't say how far you should walk, or for how long. Just, please, go for a walk.

As you walk, take note of anything and everything that comes into view that you consider to be a part of nature. You may think this a gratuitous exercise, that you already know what you'll find, but I'll wager you see something you haven't noticed before, or for which you do not have a name. Oh, and don't just look. Take note of what you hear, smell, and feel. You might even do some tasting (but don't taste any red berries or mushrooms you find). Be quite inclusive about this, but enjoy yourself. Take physical notes if you like, but mental notes are fine for now.

We'll talk again very soon.

1

Intimate and Momentous

Welcome back. The place where you just took your walk? Let us call that your *dooryard*. Its dimensions are provisional. You can enlarge it or shrink it later, as you see fit.

It is here, in your dooryard, that climate is changing. Has been changing for quite some time. Will continue to change, almost certainly at faster rates.

Such a homely term, though, "dooryard." It is still in use today in some places to describe a patch of the outdoors where the business of human work and play, in and out of doors, transacts with *the natural*, the world over which we assume we have little control. Etymologically, the word traces to New England, generally, and Maine specifically. If you've read some Walt Whitman, the nineteenth-century poet who celebrated the United States as no other, you might recall that in his elegy to Abraham Lincoln, "When Last the Lilacs in the Door-yard Bloomed," the word is hyphenated. That's fine. Whitman was from away, a New Yorker. But: dooryard. In designating a *kind* of place, the specific meaning of the word draws a contradistinction to the front yard (a space meant to impress society) and the barnyard (a place for working with animals).

The dooryard is close in, a place for work but also for exchange. It's where the kitchen garden might be, full of herbs and simple greens, as well as a few varieties of posies. But it's also full of insects and is a favored place for cats or, if there are no cats, then other creatures that make their livings by living as near to human habitations as they can— mice, for example. Or voles.

I have had a dooryard or two, but before I describe them, let me show you another. I've ventured there a couple of times, and I've encouraged students to go alone, or with a friend or two. I went with friends myself, and that was nice, but on my first visit, I went there alone, and that was ideal.

The space surrounding Henry David Thoreau's cabin at Walden Pond in the years 1844-46 was all dooryard. There was no front yard, because although he had visitors at the cabin, Henry—let's call him Henry—was unostentatious and little concerned about what visitors should find when they arrived at his door. He briefly contemplated making them sit on pumpkins, but clearly thought better of it. There were chairs in the cabin, three chairs altogether. There was also no barnyard, since he kept no animals and there was no barn—just a woodshed, which is dooryard architecture.

A visitor today knows right where the cabin stood and just where Henry placed his door, thanks to some keen archaeological work undertaken decades ago. Obelisks and graceful chains mark the cabin's walls. Using a dose of imagination, you can walk around the cabin. Or you can go inside, turn around, and see very much what Henry saw when he opened his door and peered through his dooryard.

I almost wrote "past his dooryard," but that would have been wrong. For the two years that he lived at Walden Pond, the whole landscape was his dooryard—the trees, the slope leading down to the pond, the pond itself. It stopped at the railroad tracks. Where the Fitchburg train passed, that was another place.

I stood there one November day, alone, when the sky was overcast in low clouds and the calls of crows echoed through trees, which had recently shed their leaves. Being a historian, and an academic one at that, generally unsentimental of mind, I was surprised by the thoughts I had as I gazed out through the trees and across the pond. Here I was in this place, this hallowed place. Not far from here, no more than three or four miles away, a handful of men had the moxie to begin a war of independence. And they succeeded in winning their freedom and freedom for generations that followed.

Becoming and then being free, what did they do with their freedom? The answer was under my feet. Here, an irascible twenty-something decided to see what life, shed of entailments, might be. Later, he would report his findings to the world in *Walden; or, Life in the Woods.*

Walden began life as a book about freedom, but it wandered off into a description of the woods themselves, and of the lives that filled the woods, more than about Henry's life. While there, Henry began a lifetime habit of noticing and making notes about first appearances—the first appearances of flowers, of leaves, of birds.

From his attentiveness, Henry came to believe that November was a separate season, unlike any other. Just November. That was what I experienced in Henry's dooryard: freedom, and the season of November.

Or did I? Beginning in 2002, Richard B. Primack, a botanist at Boston University who had previously ventured to places like Borneo to collect field observations, turned his attention to Walden Pond and to Thoreau's records of plants and animals—mostly plants. Primack discovered that many of the plants in Thoreau's notes could no longer be found there. Just as important, other plants still growing in Concord flowered at different times from those that Henry observed. Primack concluded that Walden—Henry's dooryard—was warming.

Thoughts of Walden bring to mind my own dooryard, for I once had the luxury of a dooryard when I relocated to a very small town in Maine (where to call the town a "village" would seem an affectation). I was nearing the age that Henry had been when he built and moved into his cabin, and my move was perhaps proportionally equivalent to his. Concord is to Walden Pond as Boston is to rural Maine. I had been born in Portland twenty-five years earlier, and though I only recalled vacations there with family, I kept a romanticized vision of it in my head, aided at that time by the many letters and essays of E. B. White, whose book of correspondence had been published just the year before.

Whether close to the coast or inland, houses in Maine that date from the 1800s once supported subsistence agriculture, at least, if not fullblown agricultural output. The one I rented for my first year there was more the former than the latter, and it saw me through four full seasons. The house itself was what is known as a two-story colonial. It had an ell, with a large kitchen below and a bedroom above. Unlike many of the houses in town, the ell was the end of the line. There were no additional connected buildings. By tradition, many houses (both in town on a few acres or less, as this house was, and farther out on the blue highways and back roads) were connected: big house, little house, back house, barn. But even without a connecting back house (the little house was the ell), the property had a full set of yards—front, barn (for there was a barn), and dooryard.

I moved in at the height of winter, within a month of meeting the inspiration and model for this adventure, E. B. White, author of *Charlotte's Web* and for several decades the voice of the *New Yorker* magazine, at his home in North Brooklin, Maine. I had knocked on his door late on a January afternoon; he famously developed a strong dislike for this sort of intrusion, but I have a gracious reply to a letter of thanks I wrote him, and I later came to make a distinction between summer visitors and winter visitors. Perhaps he did, too.

My new home felt like the edge of a Great American Wilderness. It certainly was different from any place I had lived before. One local fact that I had difficulty ignoring, for the first month or two anyway, was that the elderly woman who would have been my next-door neighbor had been murdered in her home the week before I rented. My first visitor was a detective investigating the case (I soon learned that many of the townspeople seemed to know who did it; there was never an arrest). He assured me I had little to be concerned about.

I was essentially faced with a choice: live in E. B. White's Maine or live in Stephen King's. I chose the former. White's essays were full of pointers on how to succeed at rural life down east. Like other essayists before and since, White seemed fond of winter and winter's rhythms. Reading *One Man's Meat* gave me a vivid and sensuous image of something as mundane as a late evening visit to the barn, to check on the animals. I did not yet have farm animals, but I paid visits to the barn at night as though I did. I sent for seed catalogs and purchased the minimum gear needed for raising chickens. In the meantime, doing without a car (and resenting the need for one, to which I would eventually capitulate) my world was circumscribed by the distance I could walk daily on the sides of winter roads.

The house was "in town," where the houses were densely config-

FIGURE 1.1. My dooryard at sugaring time.

ured on lots of about an acre each. There was a general store (groceries, hardware, widgets), a gas station, and another smaller store that had only recently added refrigeration. This was known as Dot's, although Dot hadn't owned it for some time. After a spell, I would pay my bill at Dot's by baking apple pies for sale; that first year I ran my bill up and paid it down as best I could. I was able to make my living, or nearly so, doing freelance work for publishers in Boston, and so needed only to walk to the post office, the general store, and Dot's.

Snow fell regularly that winter and was cleared by plow trucks that absorbed much of the town's budget. Behind Dot's and beyond the bridge leading into town there was a smelt camp—a grouping of twenty or so small shacks that had been set out on the river ice so that fisherpersons could sit inside, warmed by a small woodstove, and fish for the little fish. At night, the space between shacks was lit by a string of electric lights. With these, the smelt camp gave the town a sense of something happening, possibly even some excitement, but as White had said nothing about smelt fishing, I put off a visit to some future time.

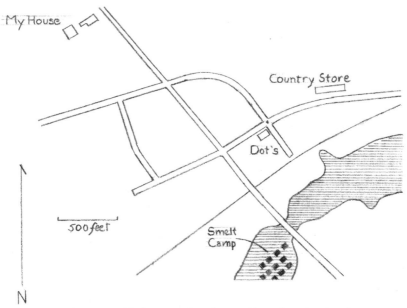

FIGURE 1.2. The geographical extent of my dooryard in Maine, showing places I mention in the text. It was not a big place.

I did order seeds and eighteen chicks. I also bought taps for the maple trees in my dooryard. The chicks came first, and on that day, the post office was filled with the sounds of chicks in piles of cartons. I raised the chicks successfully, but a marauding member of the weasel family called a fisher, for which I was ill-prepared, killed all but two of the resulting hens. I slaughtered, plucked, and ate one chicken. The other provided brown eggs for a time.

The poultry aspect of my adventure didn't go especially well, and I tell of it only out of the wish to establish myself as a reliable narrator. But the maple syruping part was a success—in my eyes at least. My dooryard was lined with maples, and when conditions were just right—warmer temperatures in the morning, freezing at night—I drilled and tapped the trees and hung gallon milk containers that I'd collected for the purpose. These were unsightly but common; galvanized pails with little rooflike covers were a rich man's decoration. I gathered and evaporated enough sap to make maple syrup for my own consumption for a year, plus some maple sugar candies. I used up a lot of propane to do it.

As days grew longer, and the snow thinner, the sap stopped flowing. The smelt shacks had already come off the river. The river being tidal (a necessary condition for smelt), the ice groaned for several weeks as it rose and fell twice daily with the tide. It was so cold that year that the Coast Guard came up the Kennebec River with an icebreaker to get things moving. Crocuses called for a coming of spring—in Maine, the coming of spring lasted from early April until sometime in June— followed by the appearance of forsythia. The spring peepers, a type of aptly named chorus frog, came and went. I knew it was summer when the side yard (another oddity of this house) was suddenly filled with day lilies in bloom. Summer, hardly my favorite of the seasons to begin with, was made worse owing to the overabundance of freelance work passed on to me by publishers, but it fortunately passed on to fall in short order. None of seeds I ordered had gotten into the ground.

Fall brought longing. Some people graduate from high school or college and never look back. Others ache to revisit texts and notebooks as soon as the first leaves change. I am the latter type. I eventually resolved that seasonal nostalgia for campus life by remaking myself as a college professor. That year, though, I delved into autumn chores. My neighbor across the street was a local history buff and organizer of the historical society, which met in October around a cider press. We made cider and started talking about preparations for the winter to come. Would we all have enough firewood to "spring out"?

As these were the "energy crisis" years (Jimmy Carter was president and Michael Dukakis, governor of that state to the south, appeared on television appropriately dressed for his lowered thermostat in a cardigan sweater), we all thought and talked about energy the way others talked about real estate a few years later. My rent that first year, reasonable even then at $275 a month, was eclipsed by the cost of heating oil in such a large, uninsulated house. By midwinter the following year, I had moved to more efficient lodgings.

I was learning. And I was not and had never aspired to the role of back-to-the-lander. After a period of atonement in a tiny apartment closer to Dot's, I lived for seven years as a tenant on an apple farm several miles from the center of town. I planted a few of the five hundred apple trees on the property and took time to draw one of the bare root trees before I put it into the ground. But that first year, thanks to the absence of an automobile through the better part of it, I was more than usually attentive to the seasons.

Seasons are a *fact of life*, and few things in life are experienced with greater intimacy.

Good records of the sort that would be helpful are the core of *phenology*, the study of seasonal change. (I have heard it pronounced with a long *e* and a short one. Either is probably correct.) Phenology is best described as a minor science that has been practiced by major figures, including Thomas Jefferson, Henry David Thoreau, and Aldo Leopold. "Data" of the phenological kind also fill the diaries and letters of American men and especially women over the past two hundred years. As agricultural colleges developed in the post–Civil War years, phenology was sometimes offered as a course of instruction. Today's climate scientists have made use of these records as a way of identifying consequences of climate change.

To practice phenology means, specifically, noticing and then recording events that occur seasonally each year: the first buds of maple leaves, the first robin in the yard, the day that fireflies appear on a summer evening. Weather events are also of phenological interest. When was the final frost of the spring? And the first in the fall? How many days did cumulus clouds appear in the sky? How often did the tornado sirens sound?

Let us return to that list I asked you to make in the prologue. It is a rough draft of a list of potentially endangered entities. *Your* endangered entities list. Some of those entities are species. But if you were inclusive enough, some of your list items are *other*: rocks, perhaps, or the webs of spiders (the webs, unlike the spiders themselves, are not species). They are endangered because of climate change. On a walk ten years hence, or twenty or thirty, a list made in the same place, on the same day of the

FIGURE 1.3. A very youthful apple tree, ready to be planted.

year, will be different. How it will differ is difficult to say. And that is much the point of this book.

Your dooryard may differ from what I just described in details small and large, and I have stretched and refashioned the term "dooryard" to mean something more than the Maine dooryard. When I write "dooryard," I am referring to your neighborhood, your environs, your surrounds.

As climate changes in an acute fashion—and it is changing, acutely—the world is also changing, your dooryard with it. The change is acute because it is in response to specific human actions that have increased by orders of magnitude over the past two hundred years, a very short time in the history of the planet. I will not try to convince you that this is so. By now, it is obvious to anyone who is paying attention. Anthropogenic climate change is an undeniable fact, albeit a *noisy* one. And it is so clearly the case, the fact of the matter, that if you don't believe it, you are practicing what the American philosopher Charles Sanders Peirce called "the method of tenacity." There is almost nothing I can say that will change your mind, and so I will not try.

That last was rhetorical. *You* already know that the world's climates are changing, you know *why* they are changing, and you have a sense of what needs to be done to contain the variety of catastrophes that we face. None of that is the subject of this book. Instead, the pages that follow are about how to live in *and how to know* a changing world. Because, have no doubt: changes are unfolding.

What those changes will be, it is difficult to say. In the broader sense, ice will go out of the Arctic Ocean, in the summer at least, and sea-lanes will open up. Cities—new Chicagos—will be built on the north coasts of Canada, Alaska, and Siberia. Polar bears, those that survive, will be drawn to the trash heaps of those cities. There will be famines in places that once were better watered. Coastal areas, working seaports and city parks, will vanish, to be replaced by new versions of their old selves, in new places.

In their zest to alert the rest of the world to the alarming fact of climate change, scientists have made some choices concerning what to talk about in order to get our attention. Some of those choices have caused confusion. But most of the confusion about climate change is intentional, as Naomi Oreskes and Erik Conway demonstrate in *Merchants of Doubt: How a Handful of Scientists Obscured the Truth on Issues from Tobacco Smoke to Global Warming*. And most of that manufactured confusion counts as malevolence.

Then, there is honest confusion. "Climate change" and "global warming": even without the handful of scientists that Oreskes and Conway talk about, the two phrases that designate this set of phenomena each contain one or more red herrings—words and concepts that scientists use in a slightly different way from the way laypeople use them. First: climate change. "Climate" is not a synonym for "weather." Weather is palpable: it's hot or cold. There are clouds or there is blue sky. It's raining, snowing, or not. Climate is different. Climate is an abstraction. We don't experience climate, in the way that scientists use the term. Instead, climate is a description of a number of factors, temperature among them, over decades. Thirty years is a customary measure. Climate is inferred from records kept over generations. It's not something to be judged from a given day's weather, or even weather in a given year.

If one cannot find climate in weather, where does one look? The simple answer is that climate is best seen in plants—botany—and change is best seen through slow, patient phenological observation. We will come back to both botany and observation.

If the disjunction between climate and today's weather is not bad enough, there is also the problem of "climate" in the singular. The earth possesses not a single climate, but many climates, each interlinked to others in complex ways. The same difficulty is found in the word "global." Climate scientists have devised ways to measure a global mean temperature, but it's not something that any *one* person experiences *as such*.

And that brings us to "warming," which implies temperature. To a nonscientist, temperature is primarily a measure of comfort whereas for a scientist it is a measure of energy in a system. Simply put, if you think of the earth as a system made up of systems, there are at any given time a large number of adjustments and adaptations to changing conditions. If the change in conditions is gradual, then the adjustments and adaptations are gradual as well. But when the change in conditions is rapid, the adjustments and adaptations are also rapid. Increase the energy in the system, and you will see a change in conditions.

Some time ago I bought a Volkswagen bus that had been modified to improve its performance. Unfortunately for me, the modification was illegal. In order to register the vehicle, I had to procure and install all of the original components that were standard in the year the vehicle was produced. I did so—it was something of an ordeal—and I returned the engine to its original configuration, or nearly so. An error in the mix of parts allowed air to leak into the system, causing what mechanics call "lean running." In time, lean running led to the demise of the engine, as heat from combustion in one cylinder led to a molten hole through one piston.

This happened on long-distance journey, as I was driving up an escarpment that rose from sea level to five thousand feet in a matter of a few miles. Would it have happened on level ground? I'll never know. But here's what happened. The hole in the piston allowed the combustion in that cylinder to force most of the oil out of the engine and onto all the surfaces in the engine compartment (this is known to mechanics as "blow by"). In a matter of seconds, this would have starved the bearing surfaces on the crankshaft of oil. But before that could happen, a light came on and I shut down the engine.

Is this a good metaphor for climate change? It is and it isn't. On one hand, my engine, like Earth, is a system made of systems. On the other, Earth is many times, indeed many magnitudes of order, more complex. Earth is not equipped with an oil light or with a driver who is committed to preventing complete catastrophe. But otherwise, for this discussion, the metaphor is apt.

There are predictable things that will happen, are in fact occurring, and ought to substitute for an oil light. Ice on land is melting and flowing into the oceans, raising sea levels. Ice that has been secure on land is slipping into the oceans and is raising sea levels with the potential to do great harm. At the same time, as the oceans absorb heat, they increase in volume, also raising sea levels. And the warming oceans are

becoming more acidic. These consequences are ongoing, and the processes behind them are somewhat predictable.

But sea-level rise, at current rates, is not simply like filling a bath-
tub. Today's shorelines have adjusted—and shore-related biology has
adapted—to a relatively stable sea level. Change the sea level and
shorelines change (adjust and adapt) in ways that aren't entirely predictable. And I've said nothing yet about the increasing frequency and
intensity of storms.

This is just one example among countless examples. We do not know
enough about the world, yet, to predict everything. There are many
changes that we will just have to learn about as they happen, and this
will be true for our children and our children's children. The consequences of climate change, in their great number, are unlikely to be all
of a kind, and few of them are knowable in advance. They are likely to
be uneven—lumpy is a fine word—in character.

Of the natural sciences that help us make sense of climate change
and its consequences, *ecology*—the study of the relations between organisms and their environment, the area most likely to be affected by
change—is among the least developed. Offering no weapons for waging
and winning wars, little in the way of pharmaceutical interest, no road-
map for finding oil or gas, ecology has had something of the status of
a Dickensian stepchild among the sciences, even as "the ecology" became a synonym for "the environment." There is much that ecologists
and their practical fellow travelers, conservation biologists, know. But
there is much that they do not know.

They are about to learn a lot, if they can gather the immense harvest of data that is potentially forthcoming. Let us call climate change,
provisionally, an experiment of an inadvertent sort: a grand experiment. Disturbances in otherwise stable systems provide opportunities
for understanding. Francis Bacon, a proponent of scientific inquiry in
the early decades of the Scientific Revolution, suggested as long ago
as 1620 that "the nature of things betrays itself more readily under the
vexations of art than in its natural freedom." What greater "vexations
of art" could there be than in rapidly ensuing changes in climates?

We stand to learn a lot. But is it ethical to gain—in this case, to gain knowledge—from what many would argue is an injustice? This is a hard question, and I cannot fully answer it here. But I can suggest that much that we learn will aid in mitigating additional destruction.

The best thing that could happen, today or next week or next year, would be to reach agreements, intranationally and internationally, to cut carbon emissions, to roll them back, and to begin mending the damage done to the planet, especially that of recent decades. *This may yet happen*, and that is what I, and what I can imagine you, hope for. As of this writing, however, it does not seem likely in the short run.

In the meantime, it seems as though it might feel morally wrong to describe cumulus clouds or count fireflies when whole nations pass below sea level. In doing so, we seem to accept climate change. Is acceptance, in this case, causative? In some sense it is, particularly if acceptance is our only response.

The range of actions that can be taken to slow climate change, and even to reverse its course, is great, and these actions fall into two primary categories: mitigation and adaptation. In the area of mitigation, individual and voluntary actions to reduce one's carbon footprint are surely worthwhile and serve as signs to others. Political action is more important to mitigation, at present. Adaptations are inevitable, even when the need for them is met with reluctance. But being aware, even while mitigating and adapting so as to ward off harm, are not mutually exclusive actions.

And so it is helpful to think of our relationship to nature in two distinct ways. We should think of human life and human action as part of nature and, yet, as distinct from and in some degree opposed to nature—both of these, at the same time. Philosophers know the second of these attitudes as *dualism*. All human actions, from this point of view, are *un*natural. The idea of dualism is sometimes called the natural attitude, not because it is an instinct but because it is deeply embedded in Western culture, especially American cultural traditions. Its source is the monotheistic traditions of the world's major Western religions, according to which humans are accorded a special place in creation by God—dominion, as one biblical passage put it.

As the historian William Cronon has argued in his finely sketched essay "The Trouble with Wilderness; or, Getting Back to the Wrong Nature," this set of ideas causes us to value different parts of the natural world with different levels of enthusiasm, sympathy, and awareness. To Cronon and others who critique the dualistic notions, humans are every bit as natural as anything else. We are animals, albeit crafty animals.

Does this absolve humankind from responsibility for climate change? For extinctions? For a path of destruction that is unfolding even now? Is *anthropogenic* climate change logically, then, simply *natural* climate change? These are difficult questions—revealing in some ways and concealing in others. If we choose, on the whole, to let climate change unfold, it will. But the *Anthropocene*, as some have taken to calling this geological epoch, could just as easily be a time when we crafty animals recognize the damage we are doing and reverse course.

The dualism/antidualism notion does not, in itself, solve problems. But it is useful for thinking about how to see the world from our dooryards, to which we again return.

Today, in New England, where I had my first dooryard experience, syruping occurs later and is threatened by climate change; the sugar maples produce sap with lower concentrations of sugar and are disappearing from parts of their former range. Smelt fishing, too, seems to be off. In some years, the ice isn't as thick for as many weeks as it was in the past, and recently, when it was, the smelt weren't there in their customary numbers. Fall arrives later than it did. Careful records of the comings and passings of seasons over the decades would provide a clearer picture of how the climate had changed, had I kept them. I possess memories, some of them true and some romanced, but little in the way of data.

In the chapters that follow, we will look in detail at phenology, its history and practitioners, and at the relationships that obtain between the sun and our tilted, orbiting planet. We will examine the drivers behind climate, the ways that climates change, and the forms that landscapes take in response to climatic factors, stable and changing. For those who appreciate no more than an exemplar and a gentle push, that may be enough. The rest of us might like a little more guidance, and this book

FIGURE 1.4. The smelt camp operated long into every evening, when fishers would take shelter in the small huts on river ice. The number of weeks—or days—when the ice is thick enough for the camp has shortened, and the smelt come at different times from what people remember from the past.

will provide it: how to establish baselines in photographs, maps, and lists. (I will advocate "kite aerial photography" and rephotography as ways of capturing baselines.) There will be guidance on how to make useful observations of weather, plants, and animals, along with explanations of how different species of animal and plant become adjusted to seasonal change and climatic equilibriums.

If you find all of this bracing, I will recommend that you practice citizen science, keeping careful records in a form that can be used by scientists and perused by other citizen scientists. If you would rather keep your own counsel, it is fine to engage phenology as a participant observer.

Nothing in the pages to follow is designed to demonstrate that our global climate is warming. That has already been amply demonstrated, and it is perverse to suppose otherwise. Neither does what follows provide a plan for turning back the clock, or turning down the thermostat. Rather, what lies ahead is a plan—partial in its scope, for there will be

much else to do—for living in a time of warming climate. Climate warming is not something to be wished for, not something to be encouraged
but, rather, being fact, something to acknowledge and to face squarely.

We learn in our science classes that scientists make observations.
They also develop hypotheses, devise and execute experiments, test
theories—so goes the litany of scientific method. But let us pause to
contemplate those observations. An observation must have about it the
characteristics of the here and now. It cannot be speculative, reaching
somehow for the future, for that would taint both the process and the
outcome. With this presentness, scientists share some part of the spiri-
tual experience of one who is mindful, someone who is attuned to what
is here, and now.

This is true no less of phenological observation than it is of any
other pursuit in science. Each observation I make—of grasshoppers
or bluebirds, of small pools of water from yesterday's rain—binds me
intimately to my dooryard and to the present, even as climate change
grinds along, like a steamroller on an endless road, momentous in all its
implications. My presentness, and yours, releases us temporarily from
the incessant, and often noisy, directionality of climate change.

2

Seasons and Circulations

When I think of it—and this episode comes to mind often enough, like a book opening to a dog-eared page—I think of it as *absolute autumn*. I've forgotten exactly where I had the experience, and cognitive science goes a step or two toward suggesting that what I *think* I remember may be a confabulation, a mash-up, of two or even more moments in memory, spread out in time. Be that as it may, I can close my eyes and see it vividly in my mind's eye, faulty as that apparatus might be. It is a misty autumn evening. A light drizzle *had been* falling, but isn't now. The air is not warm, but neither is there a chill. The change from daylight savings to standard time places me in a landscape illuminated, in early evening, by streetlights and lamplight spilling through the windows of nearby apartments. Piles of damp brown leaves cover the ground in uneven layers. Here and there the branches of a tree drip moisture onto the leaves below.

Drip.

Drip.

Did this happen in southern Ohio? Some years ago I attended a meeting there in early November, so that might well be where the image formed. Or was it in Maine, where I love to make a life whenever I can do so? Maybe it germinated from a youthful memory of an afternoon newspaper route I delivered, long ago and far away. The facts no longer matter so much. It's not as though I were on trial. What *does* matter is that it is absolute autumn, *my* absolute autumn. It's not the brilliant autumn-of-many-colors that New Englanders cherish and deign to share with flatlanders from away. It isn't the golden leaves of aspen against a cobalt blue sky that one finds on the Colorado Plateau. Nevertheless, it is my singularly favorite moment from my strongly preferred season, the season that (were I rich) I would pursue across the face of the globe, throughout the year. Never mind that the moment may be a snapshot in memory or an impressionist masterpiece (so it seems to me) that my mind has painted. I go in search of a similar window on the world, my absolute autumn, at least once each year, every year. When I can.

Seasons shape human responses to them and are often unforgiving of evasions of calendar. And yet they are ill defined in some ways, different from each other but oddly dissimilar each to its own. As such, they provide novelty. Would I really be happy with a year of absolute autumns and nothing else? Surely not. But climate change threatens a flattening of the seasons, even be it slight. Of course, climate change carries with it far greater threats, such as sea level rise and extinctions, but there are other places to read about these. Let's think about seasons and the climatic circulations that shape them and make them geographically distinct from one place to another.

Seasons come in many forms, varying within limits from one year to the next. They can indeed be quite different in differing localities. Made up of a mixture of physical characteristics—winds, temperatures, varieties of precipitation, and sounds—they also vary in their patterns of life. Reproduction, movements from here and there and back again, growth, and even idleness have seasonal signatures.

Newcomers to places like Los Angeles, San Diego, and Orange County tend to complain of the seeming lack of seasons in California. They might appreciate an orientation to the subtleties of seasonal change there. The equivalent in Southern California of my wet leaves in evening is a morning that occurs with some predictability in December, when a cool Santa Ana wind brushes loose fronds from the palms lining "surface streets" and litters the pavement with them. Spring, in the same places, overfills the local color palate for a few humid weeks in late March, when poppies and other wildflowers line usually brown hillsides.

You doubtless have a favorite season and cherished moments from that and other seasons, memories of one place or many. What are they?

And, more broadly, how much of your life is shaped by seasons? No doubt you've reflected on this before, but take a few minutes to think about it now. If you live in any of the states in the northern tier of the United States, winter is a force to be reckoned with. Snow and ice make driving difficult. There are heating costs to cover. A special wardrobe, quite unsuited to summer, takes up space on closet hooks and the backs of vacant chairs. In the southern tier, you may be challenged by arid heat from mid-Texas westward or humid heat from the Atlantic coast to mid-Texas.

And summer. It's my least favorite season, but many swear by it. By law, all of us were students at one time, and many of us recall summers as times apart from the other three seasons, a time for freedom and personal growth. If you're a gardener, summer is a time for a special and beloved kind of work, bracketed by tilling and planting in late spring and harvest or final chores in fall. If you love to ski and treasure your time on the slopes, winter is a time for weekends and vacations in mountain places.

How do seasons act as a marker of time, of our personal life rhythms? More to the point, how much of our sense of selfhood is measured out in seasons? Long-term memory seems to be an essential part of what goes into making a sense of self, although the processes of self-making and memory are complex and not a little fraught. Still, we remember places, people, interactions, books we've read, and songs we've heard. They all go toward self-making. Fragrances and odors have a special way of stirring up memories, but we don't necessarily "remember" the stimulations to our noses, only what we've associated with them. Do we remember seasons? And do they help to make us who we think we are? Recently I read an interview with an author of some note in the *New York Times*. The interviewee spoke of a favorite bookstore in Manhattan. In the strange way that words on a page misfire, taking the reader off in some direction never intended by the author, I was suddenly reminded of an image of a bookstore from my past. I didn't know which bookstore—that information was missing from the image-memory. But I knew which season it was. It was winter; it had snowed and the sky had cleared. The store was a little drafty each time a new customer opened the door. More important, it was filled with the special abundance of light that happens in winter, on a clear day following a snowfall. And, since bookstores aren't a marginal aspect of my life as a consumer but something central and essential to my sense of self, it was a seasonal memory that defines who I think I am. I have many such memories.

Please, test this for yourself. When you remember things that are important to you, of importance for defining who you think you are, what is the setting for the memory? Do you recall weather and season? Do you recall trees, or birds?

For my part, I have, thus far, spoken of seasons primarily as subjective experience, indeed as my personal experience. But how have people experienced seasons through historical time? *What* do they experience?

Groups of humans have, in the past, and some in the present, responded to seasonal change by moving—migrating—from one locale to another. It's commonplace to associate seasonal migration with hunter-gatherers from articles and pictures in *National Geographic Magazine*, or the books of one anthropologist or another, but I once had a landlord who, well into his late thirties, supplemented his annual income by heading off to another state to pick apples at harvest time, returning only when he had made sufficient income to keep a stocked larder and bank account through the winter. Like him, migrant farmworkers follow seasonal patterns. And the architectural students of the Taliesin Fellowship, the school founded by Frank Lloyd Wright when he was about seventy years old, have journeyed back and forth between Wisconsin and the Sonoran Desert in Arizona each year since the middle of the 1930s.

Without migrant workers, the food system in the United States as we know it would be impossible. But even while much about modern farm life is more fixed in place, the work to be done varies greatly through the seasons, oscillating in particular between the growing season and its aftermath. There are times for plowing, for planting, for thinning, for watering and weeding, for harvesting, and for maintaining fields, outbuildings, and equipment.

For city people, seasons are sometimes more of a nuisance than a

blessing, much less a guide to life. More than once, as a college professor, I have had to remind myself which term I was teaching. I would
look at a calendar.

Henry David Thoreau, who avoided a professorial career, was attentive to seasons and worked on an almanac as a way to revise the number
of seasons, not unlike the way that biologists revise taxonomic groups.
Henry recognized five, not four, seasons. For him, the missing fifth was
November. He embraced seasons at a time when New Englanders were
working to put as much of the business of seasonal change behind them
as they could, a relic of their peasant past. Machines driven by water-
power could cease to function by winter, but the steam engine could
drive a machine twenty-four hours a day, 365 days a year, as long as
there were workers to feed it. With the invention of cheap air conditioning (cheaper in dollars, not carbon), industries moved to the Sunbelt
and workers followed, losing the beauty and health benefits of changing seasons as a price to be paid.

Biologists are working to understand how organisms are, and have
become, tuned to respond to time cycles of many sorts. Seasonal cycles
are only one of several such cycles. The cells of many organisms are
equipped with twenty-four-hour clocks to regulate metabolic processes. Human beings have cells like that. Acting in opposition to our
built-in clocks has negative consequences, and *that* is something that
every college professor who teaches a class full of eighteen-year-olds at
8:00 in the morning knows. In contrast, researchers have found little
correlation between variations in human behaviors over time and lunar
cycles. But seasons have a profound effect on reproduction in most organisms; many species see a decline in their reproductive apparatus
during some seasons. Humans are not among them, but being born in a
particular season can have an effect on one's susceptibility to diseases,
from asthma to seasonal depression.

Whether you are attentive to seasons or feel alienated from them
by modern technologies such as air conditioning and snow tires, the
responses you make to seasons are actually adjustments to a system:
"seasons" are the brand name for a system of timely shipping channels
and distribution networks that deliver the underlying conditions for

climate—and therefore climate change—to your dooryard. The basic causes for summer, winter, autumn, and spring are quite simple. But the connections between those causes and the seasons in your dooryard are anything but. They include oceans, currents and winds, mountains and plains. These connections transform simple astronomical causes into complex ecological variations, such as the North American singularity, known to locals as monsoon, which brings moisture to the desert regions of the American Southwest and tweaks what might be a bland desert landscape into an arid paradise.

The root cause and overly simplified explanation for seasons is astronomical. Our blue planet tilts on its axis of rotation (see fig. 2.1), an imaginary line, a kind of imaginary axle really, that passes through the north and south poles. The axis of rotation stubbornly points to roughly the same place in the universe no matter where the earth is in its annual orbit around the sun.[1] On the north end, the axis points to the vicinity of the star Polaris, in the constellation Ursa Minor, or the end of the handle in the Little Dipper. It is thanks to this obstinate orientation that we have seasons. When the axis at its north end points away from the sun, the Northern Hemisphere has winter, because less of the spherical surface of the earth is exposed to sunlight in the north each day (and none of the highest latitudes see sunlight at all). At the same time, the Southern Hemisphere is having the opposite exposure. Six months later, all of this reverses.

This happens like clockwork, so to speak. Astronomical spring follows winter, and summer follows spring. Calendars predict the exact moments of the vernal and autumnal equinox and the summer and winter solstice. With a few tools and some record keeping, not only can you tell where you are in the course of the astronomical year but you can also—with some additional knowledge and a couple more tools—figure out where you are anywhere on Earth, with an amazing degree of precision (as long as it's not cloudy). The power of astronomical sea-

---

1 This is an oversimplification. The point at which the axis points actually passes through a thirteen thousand-year cycle. But this has no bearing on anthropogenic climate change, and people who suggest that it does are prevaricating.

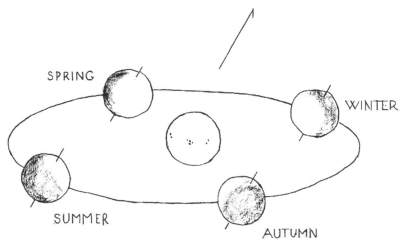

FIGURE 2.1. This, in pictorial form, is the astronomical explanation of seasons in the Northern Hemisphere. The *arrow* points to celestial north (as oriented from the earth), and the earth itself spins on an axis pointed close to celestial north. (The North Star, Polaris, is nearby.) As the earth orbits the sun (*center*), this obstinate orientation creates the seasons, as in the summer for the northern Hemisphere, when that hemisphere has more exposure to the sun over the course of a day than it does it the winter. Of course, this drawing is preposterously far from being "to scale," and the explanation for seasons as we experience them is quite a bit more complex than this.

sons to tell us things we need to know is profound, and that's part of the reason that every educated person tends to know the names Copernicus, Galileo, and Newton.

But seasons, as we *experience* them, seem to be variations on a theme rather than a simple repeating melody. Winter almost never comes to an end on March 21, and it is quite rare that June 21 is the hottest day of the year. Our experiences don't seem to match what one and only one of the sciences tells us is so. Calendrical seasons are conventional divisions of the year, with anchors in astronomical reality, but also show a preference, a bias really, for a particular kind of symmetry that makes special days of the equinoxes, out of all proportion to their significance.

In fact, the stuff of seasons is a fluctuation, an *oscillation*, from summer solstice to winter solstice, and then back again. This fluctuation is best illustrated by the sawtooth pattern of the Keeling Curve (a graph plotting changes in atmospheric carbon dioxide concentration), which shows rising carbon dioxide that peaks as the winter solstice arrives,

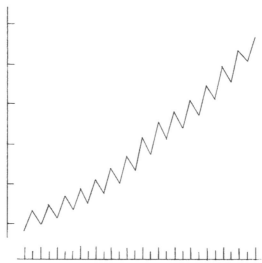

FIGURE 2.2. An idealization of the Keeling Curve, an iconic image in climate science. The x-axis is time, in six-month periods (the larger ticks are years). The y-axis is carbon dioxide in the atmosphere, measured in parts per million. The specific quantities do not matter here, as they are fairly consistent over a period of more than fifty years and are widely available. What matters is the consistent rise in carbon dioxide from year to year and the annual pattern of ebb and flow between solstices as more carbon dioxide is taken up by plants during the growing season and less during the winter.

then drops as the Northern Hemisphere (with its abundant land masses and the plants living there) greens up, and, finally, falls to a trough in the time of summer solstice, when plants are absorbing carbon dioxide rather than giving it off. In *An Inconvenient Truth*, Al Gore painted a picture of the planet breathing in (Gore took a breath in) and out. Peak and trough. Two seasons, one might say. But what of the other two seasons?

Although they do not go so far as to call them seasons, ecologists and plant biologists recognize four periods of biological activity in each year. They call these *phenophases*: the *green-up*, when deciduous plants come out of dormancy and unfold leaves to get busy with photosynthesis; *maturity*, when the leaves turn their maximum area to the sun; *senescence*, when leaves cease photosynthesis; and *dormancy*, when there is no photosynthesis and, on many plants, no leaves at all. While these phases occur in an order identical to spring-summer-fall-winter, each phase is at best an indirect response to the position of Earth in its annual orbit (length of day does matter, though).

Phenology is closely related in its subject matter to chronobiology (not to be confused with biochronology, which is a branch of paleontology), which is the study of the mechanisms in organisms that take cues from astronomical cycles—days and nights, seasons, and years, although the latter are simply expressions of seasons—as well as other cycles, such the sleep cycle in humans. Among the most powerful of the chronobiological cues in phenology appears to be photoperiod, or length of day (meaning length of daylight). This seems to be the only cue for some organisms, the most important for others, and one of two or more for yet others. Response to photoperiod is programmed in an organism's genes.

But temperature is important.

In 2016, scientists and policy makers met in Paris to try to craft an agreement between nations that will keep global temperatures from increasing above two degrees Celsius over a baseline temperature. Many thousands of people are involved, and (serious) news is filled with accounts of the problems and difficulties of achieving agreement and compliance. This has happened before, in Copenhagen, in Kyoto, in Rio de Janeiro. It will happen again. And it will come down to temperature.

Consider the thermometer.

No one (perhaps a hyperrationalist, but no one else) chooses to set aside work and play based solely on what a thermometer says. Think of the last time you were sick, with a fever, let's say, and a temperature of 102.5 degrees Fahrenheit. Did you decide to stay home from work or school based only on your temperature? Or did the reading from a thermometer *confirm* what you were already feeling, which may have been sensations of fatigue, a bit of dizziness perhaps, and a feeling that you couldn't get warm enough? Did the thermometer cause you to change your behavior? Or was it merely a confirmation of an abundance of evidence that you were sick? The thermometer has been around for fewer than five hundred years, but humans have been responding to signs of temperature, inside their bodies and in the air around them, for as long as our species has existed.

Still, thermometers are handy. They confirm the illnesses we feel by providing objective certainty. No one argues with the datum that one's body temperature is 103 degrees Fahrenheit. Thermometers reassure us that we aren't crazy, or wimpy, or lazy. It really is ten below zero outside. But in a dispute between the thermometer and our visceral sense of things, we often go with our guts. Even NASA engineers decided to override decision-making protocols, based on temperature readings, and launch the space shuttle Challenger. In that case, the consequences were disastrous and tragic. So even the people most likely to give credence to the thermometer over all other inputs didn't do so.

In the case of climate change, climate scientists, activists, and some policymakers ask us to do what we almost never do in other aspects of our lives. They ask as to use a fairly fine-grained reading of an instrument as an indicator of trouble ahead. Temperatures are rising, they warn, and will rise even more.

Let's think like a scientist a little bit, so as to give them the benefit of the doubt.

As I've said before, temperature has a different meaning for a physical scientist than it does for us. A scientist thinks of temperature as a measure of energy. For example, most physical scientists have seen a demonstration of the effect of the Curie point on magnetic properties. The Curie point is a phenomenon that demonstrates what we have come to know as a "tipping point," a moment of dramatic change in the properties and behaviors of the things and people around us. The magnetic properties of iron objects change at the Curie point, and the change appears to be instantaneous. A magnet will attract a nail made of steel at a temperature below the Curie point but will not attract it at or above. Watching a demonstration of this (some museums have a Curie point exhibit) is edifying. It can leave an impression.

But a much more mundane example of scientific thinking about temperature has to do with fairly instantaneous changes in the properties of many things, the most common of which is water. Physical scientists call these changes "phase changes." Below 32 degrees Fahrenheit (0 degrees Celsius), water, at sea level (this has to do with air pressure), is a solid. Between 32 degrees and 212 degrees Fahrenheit (0–100 degrees Celsius), water is liquid. And above that, it's a gas. You can watch this phase change anytime you wish to by placing a pan of water on a stove and watching bubbles form on the bottom of the pan. Those bubbles are water, suddenly in its gaseous phase, expanding in the surrounding liquid and then rising up and out of the pan as steam. That gaseous phase carries with it the energy that you have transferred to it from the gas or electric burner under the pan. How energetic is it? Why, energetic enough to power the entire Industrial Revolution.

That's worth thinking about. Most of our modern world, from the steam locomotive to factories to electrical generation, depends on this phase change happening with absolute predictability under a variety of conditions. Liquid to gas, gas to liquid, predictably, at 212 degrees Fahrenheit, at sea level.

But here's the difficulty with all of this. Just as we don't need a thermometer to know we don't feel well, so, too, is a thermometer far from necessary for boiling a pan of water. "Bring to a roiling boil," say most recipes, not "warm to 212 degrees" (although for legal and other reasons, many recipes ask us to heat meat to specific temperatures, and I imagine that some people actually do this). Another way of putting the matter is to say that the relation among heat (the energy active in a particular pocket of matter), temperature (a measurement of that heat), and phenomena (the stuff that the energy makes happen) is most obvious to us when we look at the phenomena.

In the case of the climate crisis, phenomena are a lagging indicator, and an especially slow and inconclusive one at that. So scientists shout "temperature, temperature, temperature!" to get our attention. And they are certainly right to do so, given what they know.

But it hasn't been enough, in part because our experience sends us confounding messages.

Like temperature, astronomy is a precise and yet, at the same time, a surprisingly crude guide to how life on Earth unfolds and fares in any given year, as anyone who looks forward to the coming years *Farmer's Almanac* seems to understand. There is often much variation between one summer and the next, but anthropogenic climate change is giving that set of variations a push in one direction—warmer.

El Niño, a major contributor to annual variation in seasonal conditions in some years, is a good example. While the causes of the formation of El Niño conditions certainly derive from the energy of the sun (and involve feedback mechanisms), there is currently no theory that ties El Niño to Earth's annual orbit in any clear and predictable way.

So, seasons have a simple cause that manifests in complex ways. How do they work? Where do they come from? Why do they seem to vary as much as they do from one year to the next, and yet blur together in memory?

Seasons are the avatar of climate—a rhythmically pulsing shadow of the key abstraction in our lives and fates as it makes its transformations of matter, energy, and time. If astronomy were all that mattered, seasons would be much the same from year to year, and you could expect a close correlation between day length and temperature. But other things *do* matter—in particular, the oceans, the atmosphere, and the interactions between them. One of the first scientists to think carefully about these interactions was the Norwegian physicist-turned-meteorologist Vilhelm Bjerknes, who pictured the atmosphere in the Northern Hemisphere as a war between a polar air mass and a temperate one, each seeking to advance and conquer the other front (he took the idea of a front from the battle lines of the First World War). Over time, Bjerknes's idea has been refined, but Bjerknes was one of the first to think about local weather and seasons as a product of global-scale forces.

As the focus of this book is about your dooryard, and climatic changes that you can and will witness if you pay attention, it is important to note that the weather that delivers your *climate* (average temperatures, for instance, over a thirty-year period) makes its way to your dooryard, hour by hour and day by day, under the influence of patterns (the usual and expected), perturbations (unusual or at least unpredictable), and events at some distance from your dooryard. It isn't absolutely essential that you know or understand these patterns, but it's more interesting to know that a migratory bird arrives in your dooryard because of high pressure thousands of miles away. More to the point, all of our planet's atmospheric patterns, as well as ocean currents, are likely to change over the coming decades. How significant the changes will be is something we can only model now but will learn in time.

Phenologists have already learned that a key driver of some organisms' responses to climate has to do with the combination of day length and temperature. (There are species for which this isn't so.) As climates change, the green-up occurs earlier in the Northern Hemisphere, even though days are not getting any longer. And so while day length and temperature rise are related to each other in a causal way—that is, longer days actually do cause warmer temperatures—the relationship (or correlation) is not simple.

The length of the day, from sunrise to sunset—longest at the summer solstice and shortest at the winter solstice—has an effect on air temperatures (and on ground and ocean temperatures as well). But how high in the sky the sun is over the course of a day also makes a difference. For the higher in the sky the sun is, the less atmosphere its radiation must get through in order to reach ground level. At the higher latitudes, near the poles, sunlight may shine most of the day (or all day long above the arctic and Antarctic circles in summer), but at a rather low angle compared to the angle at the equator. So while day length is a significant driver of changing seasons, its effect on specific places—like your dooryard—is as much a consequence of geography (your latitude) as it is of astronomy (where the axis happens to be pointing).

The atmosphere close to the ground is, on average, warmer near the equator than it is near the poles. Of the two, cold (at least in a relative sense) is the default condition. To become warmer, sunlight needs to be added. Take it away, and the warmth will spread to colder places. This is the first law of thermodynamics, and there doesn't seem to be any way around it. The total energy will remain constant, but energy seems to want to be fair, to spread itself around evenly.

The thing to be treasured about the earth's atmosphere is that it makes a noble attempt to get around the first law, to keep solar energy from radiating back into space as soon as sunlight stops striking it. The atmosphere has been described as a blanket, and so it is, but it's more than that. It doesn't simply trap some of the sun's warmth, even after the sun sets. It also gathers up some of that warmth and reradiates back to Earth. This is why, for instance, a cloudy evening—in the absence of wind—is always warmer that a cloudless evening. The clouds aren't just an insulator. They are also heat storage devices.

The earth has a budget for energy, and over time it's a fairly balanced budget. Solar energy comes in, reflected and radiated energy goes out. When energy comes in, it may do some work, but it's still energy. If the same amount of energy that comes in goes back out, the budget is balanced, and the earth's temperature remains the same. If something upsets the balance, the temperature of the earth will rise (or fall) until the exchange achieves a balance once more. More about this a little later.

It's now that the story grows rather complicated—too complicated to do it justice here. Suffice to say that differences in the temperature of air from one place to another will tend to form convection cells. Differences in humidity will cause differences in the density of the air in one place relative to another. The earth's rotation creates the Coriolis effect, which changes the direction in which the air is traveling, just a bit. All of this gives rise to trade winds and the jet stream. And more. Winds passing over the ocean in a consistent way drive ocean currents, which carry warm water from, say, the Caribbean Ocean to higher latitudes. It's all quite interesting but takes a book as long or longer than this one to gather up all the details and make sense of them. The important thing is that cold air in your dooryard probably lost heat someplace else, perhaps far away, and the same is true for warm air—it may have warmed up in the area around your dooryard, but in many cases it didn't.

And if this weren't enough, there are patterns in the atmosphere or the oceans that vary, back and forth, over some period of time (the time periods vary) between two contrasting conditions. Climatologists call these oscillations.

An example is the North Atlantic Oscillation (NAO), which consists of high pressure over the region of the Atlantic Ocean near the Azores, and low pressure over Iceland (see fig. 2.3). The exact positions of these areas are not fixed; they can vary. If both the high and the low pressure

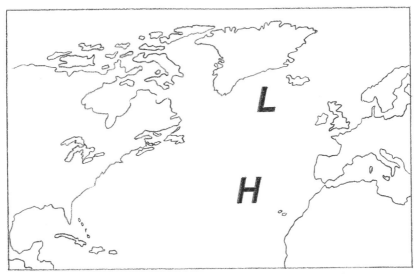

FIGURE 2.3. Typical pattern of the North Atlantic Oscillation. In its positive phase, both the low-pressure area (*L*) and the high-pressure area (*H*) are strong, with associated counterclockwise winds and clockwise winds, respectively. These hasten westerly winds across the Atlantic Ocean, and the East Coast of the United States has a warm weather pattern. Northern Europe is warm and damp. In the negative phase of the oscillation, these pressure areas weaken, the winds associated with them slow, and the westerly winds wind their way north around the low, bringing cooler temperatures to Europe. The oscillation could also be called a toggle, back and forth.

are strong, the clockwise winds of the high and the counterclockwise winds of the low combine to reinforce and strengthen the westerly trade winds. Meteorologists call this a positive NAO; under such conditions, cold air drains off of the higher latitudes in North America and doesn't build up. In the opposite case, the negative NAO, the westerlies are slowed, and cold air does drain from the higher latitudes to make the East Coast of the United States colder. If a negative NAO combines with a negative Arctic Oscillation, the East Coast will get very cold indeed.

The Arctic Oscillation is low pressure over the Arctic Ocean. It varies between stronger and weaker, like a spring. When the Arctic Oscillation is weak, it generates weaker than average westerly winds, which allow cold, polar air to spread southward. A strong Arctic Oscillation, generating strong westerlies, will keep cold air trapped in the polar region. But a weak Arctic Oscillation combined with a negative NAO will draw a blast of cold polar air to the Eastern seaboard of the United States.

Far away, the North American singularity, which I mentioned earlier, sometimes called monsoon in the American Southwest, is another pattern with fairly predictable results. The North American singularity draws moisture from Mexico and spreads it in a plume of humidity, northward on a path that follows Interstate 17 in Arizona. The pattern is annual, and predictable. Residents of Tucson and Phoenix look forward to strong summer rains, with thunderstorms, beginning around July 1 and continuing through August. The North American singularity has altered the desert ecology there. Desert plants are more numerous and more dense than they are in nearby desert areas unaffected by the monsoonal rains. The plume continues northward to water the southern Colorado Plateau, including the continuous forest of ponderosa pines that grow above six thousand feet there. The monsoon also diminishes the threat of wildfire on the plateau, and citizens of Flagstaff, Williams, and other towns there breathe a collective sigh of relief when the rains kick in.

And then there's El Niño, the best known of the oscillations. Climatologists aren't as clear as they'd like to be about the specific conditions that make for an El Niño, and so aren't able to predict El Niño years in advance. But the set of conditions that make up the oscillation are quite well understood. Trade winds in the Pacific Ocean reverse, bringing warmer water to the eastern Pacific off the coast of equatorial South and Central America. When this happens, other patterns in the region are perturbed, California gets a shot of winter precipitation, and the Eastern seaboard of the United States often experiences a warmer winter season. With El Niño, though, it is still difficult to predict the weather in the United States much more than a week or so into the future. As I write this, in December 2015, there are places where plants are well into their green-up, something that should not happen for another two to three months, at the earliest. How much this present El Niño has been a monster (wags are calling it Godzilla) and how much it has been amplified by anthropogenic climate change is something climatologists will argue and discuss for years to come.

Everything I've said thus far, about seasons and the systems that distribute heat (and cold) from one place on Earth to another, pertains to the way things were before humans started to burn coal and oil to do work. It's time to add fossil fuels to the fire.

When some factor (such as an increase in carbon dioxide in the atmosphere) throws a monkey wrench into the earth's energy budget, climatologists call that monkey wrench a climate-forcing agent. It's an odd and not especially charming technical phrase (what were they thinking?), and I'll be rid of it here as quickly as I can, but for the moment, we're stuck with it. In the meantime, it gets worse: climatologists shorten the phrase to "climate forcing" or even just "forcing" and measure it as a "radiative forcing."

Carbon dioxide is a forcing agent. So are methane and the other greenhouse gases. There are others still, such as aerosols—nongaseous airborne matter such as soot and dust. What is important about all of these is that they can alter the energy budget—*force* it out of balance and on the path to a new normal, a new average temperature. It's carbon dioxide, though, that is the primary culprit in anthropogenic climate change. That's the gas that animals exhale as part of metabolism but which plants take in. Carbon dioxide is a significant climate-forcing agent, and it increases in concentration in our atmosphere with each passing year. This was the finding of Charles Keeling, of the Scripps Institution of Oceanography, who began measuring concentrations of carbon dioxide in the 1950s. The measurements have continued to the present. When graphed, they exhibit the now famous rising sawtooth pattern of the Keeling Curve back in figure 2.2.

Just as the other greenhouse gases do, carbon dioxide absorbs solar energy and then reradiates it, since it's not just the earth but also every molecule on and above the earth that follows the energy budget. Every carbon dioxide molecule absorbs solar energy, either directly from the sun or from other reflected or radiated source, and then eventually radiates it. In, and out.

There are other "forcings." Some have increased in the atmosphere (methane is an example) because of human activity. In other cases, such as volcanic ash, humans are blameless. But methane is an important example, because it is not just a forcing agent in its own right. It's also a good example of a feedback. Before moving on to feedbacks, however, let me summarize things to this point. On the one hand, there is the earth's energy budget, which has responded to forcings throughout the history of the planet, adjusting temperatures downward (giving us ice ages) and upward (the Cretaceous Period was warmer than the present). On the other hand, there is human action, which stirs the pot—adding energy to the budget. As these forcings do their business, some of them bring about changes that will sometimes reinforce and sometimes damper the ongoing change. For example, temperature increases in the arctic cause polar ice to melt. Ice is great at reflecting solar energy back out into space. When it melts, it reveals ocean water, which absorbs solar energy. Climatologists call this a feedback process. Feedbacks can be of two generally kinds: positive feedbacks, which reinforce warming, and negative feedbacks, which diminish it.

An example of a positive feedback is when higher temperatures melt permafrost in subarctic terrain, releasing the trapped methane. Another positive feedback is increased evaporation, which can add water vapor (yet another greenhouse gas) to the atmosphere, and water vapor enhances warming even more than carbon dioxide does. But more water vapor, when it combines with atmospheric dust to form clouds, may increase albedo (i.e., the reflective power of the water vapor), reflecting solar energy and constituting a negative feedback.

It's like playing three-dimensional chess or trying to satisfy all your family members at the same time. If climate change were to do nothing more than to increase temperatures uniformly by fewer than a few degrees, this alone would be enough to change landscapes and the nature in our dooryards. (Some would refer to this as linear change, but I don't find the description helpful, so I'll drop it here.) The change would be distributed across the seasons by the atmospheric engine. Spring would occur sooner, fall would be later, and summers would be longer and winters shorter. Plants would begin their annual cycle of phenophases early and achieve senescence later (if at all, in some places).

But from the time that climate change started to worry climatologists, they began to a have concerns about the distribution system itself. Best known among their concerns is the worry that Greenland's glaciers will melt rapidly, bringing a reservoir of fresh water to the North

Atlantic, where the changed salinity (and density) would alter the system of currents in that region, or even shut them down. Were that to happen, the earth would no longer experience a uniform warming but would undergo a more drastic change. Such a change might be sudden (the direst of threats) or more gradual, over one or more decades.

The point of all this is that the principle driver of weather, which, when averaged over some period like thirty years, we call climate, does not simply warm one, two, or three degrees Celsius. If it did, there would be rather less to worry about (although were would still be trouble in River City). Instead, changes caused by warming will revise and reconstitute patterns of circulation, rejigger oscillations, put feedbacks in motion here and there. Climate does not change uniformly, or even gradually. As it changes, the seasons—not the astronomical kind, but the felt sort, what you find in your dooryard—are changing with it. Climate modelers are more than happy to roll out predictive curves of several sorts, to show how climate change could unfold. What they cannot do is to provide certainty about uniform versus not uniform or gradual versus drastic, and they cannot be very precise about what happens where you live.

It can still be said that to everything there is a season. But plan for seasons to come at different times, in different ways. The predictive climate models that scientists make will be revised by what happens in real time, as predicted temperatures are compared to real temperatures taken from real thermometers on the ground. Ground truth, this is called. Those temperatures will be averaged over a thirty-year period to provide a new climate normal. And it will be a warmer normal. As I have argued here, this doesn't tell us much about the world around us. But your phenological observations will. Your phenological observations will provide a fuller picture, in one place—more ground truth—of climatic changes that stem from carbon-forced warming.

3

Lilacs and Passing Time

Phenology, or something like it, is doubtless old as desert dust. Ancient hunter-gatherers, camped together in bands of about twenty adults, plus children, must have known their plants and by knowing them also knew where they were camping, and in what season. Somewhat later, long-forgotten tillers of the land must also have known the first appearances of each bird and flower bloom and what to make of these phenomena. Both hunter-gatherers and tillers of the soil taught their young. Or didn't, as the case may be. Those who did so prepared the generation that followed to survive, at least, and perhaps to thrive.

It's all surprisingly obscure. As I write this, not more than one in ten historians of science seems to know the term "phenology" ("Do you mean phrenology?" they ask me) and so have not dug deep in the dust to recover its history.[1] Much less have they gotten together at conferences with anthropologists, archaeologists, paleoethnobotanists, paleopalynologists, or anyone else who might shed light on the question. It's not yet a question they know to ask. And so, for the time being, the history of phenology has about it the scent of mystery. Most of the focus has been on one enlightened president of the United States; one New England walker known for his ironic writings and irascible character; one midwestern hunter turned conservationist; one English family who kept records through generations; and, more recently—too recent for historians but not for sociologists of science—thousands, per-

---

1 To someone who is not a historian of science, the field might seem small and arcane—small enough that any given historian of science must surely know pretty much all there is to know about . . . the history of science. But those inside the field know that its subject matter is vast. Essentially, it's everything that counts as science fifty years ago, one hundred years ago, two hundred years ago, and so on, plus everything that every Greek philosopher ever thought about. That's a lot to cover, and nobody can do it all. No one even tries.

haps tens of thousands, of volunteer observers across the United States
(and Europe, too, although that's another story) quietly paying attention to the universe in their dooryards and reporting their observations
through networks. They, these "citizen scientists" as they are called,
know what phenology is and are edified by the small bits of history that
they know. More important: they are excited, even enchanted by the
practice itself.

There are examples of practices that are plainly phenology, called
by that name, and there are allusions to phenological knowledge that
don't invoke the science. Here, for instance, Robert Herrick alluded to
the practical aspects of phenological knowledge in his poem, "To the
Virgins, to Make Much of Time":

> Gather ye rosebuds while ye may,
> Old Time is still a-flying:
> And this same flower that smiles to-day
> To-morrow will be dying.

Before there was phenology, there were people who practiced something much closer to the science without calling it by name. Among
them, a family in Norwich, Norfolk, in England, which is surrounded
by much the same crazy patchworks of irregularly bounded landhold-
ings that look so otherworldly to Americans landing at Gatwick Air-
port near London. Norwich lies on the River Yare, the mouth of which
is Yarmouth, on the North Sea. Surrounding towns include Cringleford
and Bixley, Rackheath and Taverham. North of Norwich, on holdings
in Stratton Strawless, five generations of the Marshams kept the oldest
continuous phenological observations recorded in English. Robert
Marsham (a Fellow of the Royal Society) got matters rolling in 1736
with notations about leafing of trees—sycamore, elm, birch, chestnut,
and nine others. Marsham noted blooms (snowdrop, wood anemone,
hawthorn, and turnip) and migrations of four kinds of birds. He was
attentive to the dates when yellow butterflies appeared (mid-February
in 1750) and the dates that unspecified toads and frogs began to croak.

The Royal Society of London published Robert Marsham's records in 1789, following the patriarch's death in 1787. That would have been a landmark in the history of phenology, but four generations of Marsham continued the practice: son, grandson, great-grandson, and great-great-grandson, on into the 1940s.

Even as Marsham kept up his observations, a youthful Thomas Jefferson wrote, in a crisp new notebook, "Purple hyacinth begins to bloom." This was March 30, 1766, a Thursday. The place was Shadwell, the Jefferson family home in Albemarle County, Virginia. No interpretation of the fact accompanies the observation, no reflection on its significance. Just five words that begin a columnar march of such jottings in Jefferson's "Garden Book," the record of occurrences in the natural world and actions in his agricultural life that Jefferson kept from 1766 until 1824, the year he died.

Whence that blooming hyacinth? It was likely planted by a slave belonging to Jefferson's father, Peter, after someone in the family, possibly Peter Jefferson himself, purchased bulbs. Hyacinths were among the successful Dutch commercial cultivars of the seventeenth century, along with tulips. Judging from today's map of hardiness zones, the purple hyacinth was an ideal garden plant for all the American colonies, Massachusetts (Maine was part of the Massachusetts Colony) to Georgia.

Jefferson was not a phenologist, although he doubtless would have liked to be, given his broad interests in natural history. Whether he was "doing phenology" is a historian's quandary.

"Phenology" is a not especially new word for what is likely to be among the oldest examples of human knowledge. A Belgian botanist, Charles François Antoine Morren, coined the term in 1849: "pheno" (Greek for "the appearance of") and "logos" (knowledge or word, study of). Unfortunately, the term doesn't seem to have captured the popular imagination in its time or later. In fact, it was probably much too close to the name of a trend within psychological studies based on the topographies of human skulls called "phrenology," a word with an altogether different Greek root. Alas, the two words sound very similar.

Today, search engines may return searches for "phenology" with the annoying question "did you mean *phrenology?*" I never mean "phrenology."

As is the case with the names of other sciences, "phenology" refers both to the study of some aspect of the natural world and to the aspect of the world studied. Thus, a botanist might say of a particular plant species—the ocotillo of the Sonoran Desert, for instance—that it has an interesting phenology. Indeed, the ocotillo has adapted to arid conditions by limiting the energy it devotes to photosynthesis and growth, as well as the amount of water it passes back into the atmosphere, by growing leaves only when there is rainfall, and shedding those leaves a short time later. Much of the year, the ocotillo is dormant and appears to be dead as a stick. That is part of its phenology.

Henry David Thoreau, the author of *Walden; or, Life in the Woods*, as well as the influential essay "Civil Disobedience" (which inspired both Gandhi and Martin Luther King Jr.) was better known to fellow townspeople in Concord, Massachusetts, as "the woods burner" for having accidently started a wildfire, charring dozens of acres of Concord woodlots. That youthful error notwithstanding, Henry became a founding figure in what, a century after his death, became known as environmentalism. Born to a family of entrepreneurs—their primary product was pencils—Henry hoped to begin a private school along with his brother as headmaster. When his brother died of lockjaw, Henry was on his own. After graduating from Harvard College in 1837, Henry joined the transcendental movement, whose nominal leader, Ralph Waldo Emerson, mentored Henry, making him almost a family member in the Emerson household. It was to Emerson's land on the shore of Walden Pond that Henry repaired in order to build his cabin and spend two years "in the Woods."

Henry's cabin and accompanying dooryard were not in wilderness, nor did Henry think of Walden Pond, much less proclaim it, as such. It was, simply, an alternative to an ordinary farmstead for a twenty-seven-year-old man with a disinterest in women or marriage, as well as a broadly inquiring mien unsuited to regular academic life. As he fully

FIGURE 3.1. "I would rather sit on
a pumpkin and have it all to myself,
than be crowded on a velvet cushion."
Henry Thoreau took unconventional
positions in *Walden,* but his fascinations
with nature have a very modern ring to
environmentalists' ears.

matured, Henry grew increasingly interested in executing a systematic
study of the natural world, even if he did not care to think of it as such
at first—a study grounded in vigorous, and increasingly rigorous, field-
work. He lived simply, both during the Walden years and later, without
need for luxuries or pretense. By doing so, Henry was able to support
himself as a sort of freelance surveyor (as most surveyors of the time
were), devoting the rest of working time to long hikes in daylight and,
at nighttime, observing the natural world, learning the names of plants,
and inquiring of those with such knowledge about the way that Native
Americans had used them.

Henry remarked, ironically, that he was well traveled in Concord. He
had opportunities to travel farther afield over the course of his life, and
he took advantage of some of them. He spent six months in New York

City, trying somewhat unsuccessfully to make his fortune as a writer. There was the trip down the Concord and Merrimack Rivers with his brother, and the resulting book by the same name. He spent time on Cape Cod and in Maine; he traveled as far west as Michigan. But that was it. One is left to wonder what he would have made of Yosemite Valley had he seen it, or Yellowstone or the Grand Canyon. Against all the untraveled miles across the United States and abroad, Henry walked the woods and meadows of Concord well, studying them in depth, knowing them profoundly.

Had he lived past the forty-four years and change that were his lot (he died of tuberculosis in 1862), Henry would almost certainly be known today as a scientist and quite likely a founding figure in the science of ecology. There are two historical reasons to support this venture into the contrary-to-fact: first, science in America itself changed after the Civil War in ways that would have made Thoreau feel welcome in its ranks. Second, Henry had already, by 1862, amassed a wealth of observations as well as raw theoretical insights. He was one of the first Americans to read *On the Origin of Species* and seemed eager to apply Darwinian thought to his own studies of landscape.

Two primary research interests show up in Henry's notebooks of the 1850s and early 1860s. One of them was what we now know as ecological succession (see chap. 5 of this book). The other was an investigation of seasonal change, or phenology, although he didn't call it that. Henry's interest was in seasons themselves.

So insightful were Thoreau's ecological ideas that it is a worthwhile exercise to imagine scrapping much of what actually happened in ecology and replacing it with Thoreauvian ecology, the science that Henry would have pioneered had he lived longer. But one need not imagine the roots of his ideas, as he devoted himself to setting them out in language so polished that one scholar remarked "it is unnatural to write so well every day."

Henry was a moralist, for certain, and warms the hearts of some present-day environmentalists while embarrassing others. Take this reflection, recorded in his journal entry for November 1, 1853:

Can he Who has only discovered the value of whale-bone and whale-oil be said to have discovered the true uses of the whale? Can he who slays the elephant for his ivory be said to have seen the elephant? No these are petty & accidental uses. Just as if a stronger race were to kill us in order to make buttons and flageolets of our bones, and then prate of the usefulness of a man. For everything may serve a lower as well as a higher use.

Thoreau used irony as a rhetorical tool, but he was often this straightforward—and as full of empathy—in his journal. Although scholars usually look askance at efforts to read Henry's work through modern eyes, this and many other passages in Henry's journals feel as up-to-date as a Facebook posting. Or perhaps something a little earlier than today. For those who know the reference, doesn't the passage above sound like the pitch for the "To Serve Man" episode from *The Twilight Zone*? More to our present point is the following, selected from a following page, noted on November 2, 1853:

What are those sparrows in loose flocks which I have seen two or three weeks, —some this afternoon on the railroad causeway, —with small heads and rather long necks in proportion to body, which is longish and slender, yellowish-white or olivaceous breast, striped with dark, ashy sides of neck, whitish over and beneath the eye, and some white observed in tail when they fly? I think a dark bill and legs They utter a peculiar note, not heard here at other seasons, *somewhat* like the linarias a sort of shuffling or chuckling *tche-tche-tche-tche*, quickly uttered. Can they be the grass-bird?

Whenever he could, Henry would find identification for a bird or a plant from detailed descriptions like this and continue to keep phenological notes as events presented themselves. Altogether, there are records for scores of plants in Henry's notebooks, making his records broad in terms of species and short in terms of time, whereas the Marshams' have narrower species counts over a far longer period of time.

Henry's notes nevertheless provide a benchmark of reliable phenological knowledge, one that botanist Richard E. Primack of Boston University was able to use in order to show the effects of climate change in at
least one location, and with startling results.

Conservation biology, whether one is a botanist like Primack or one
studies wildlife, can be a ticket to exotic and distant places. Primack
made the most of his ticket for more than two decades, traveling an-
nually to Malaysia for field studies. But then, in a move that seemed
to confound his colleagues at the time he made it, he ended his exotic
travel and looked for signs of climate change closer to home. He found
them in the journals that Henry kept, which recorded, among other
things, the date each year when Walden Pond became free of ice. It took
some work for Primack to transcribe the records—Henry's handwriting
is a challenge to read.

In time, though, Primack was able to show that plants now flower
earlier than they did in the 1850s, a difference measured not in mere
days but in weeks—three weeks, to be exact. The time until ice-out is
more variable but is invariably much earlier than it was when Henry
stood in his dooryard and looked out over the pond. In addition to a
large number of papers in scientific journals, Primack published *Walden
Warming* as a way of joining the chorus that was sounding the alert: our
planet is warming, Walden pond in step with the rest.

Is it mere coincidence that the author of the Declaration of Independence and the author of "Civil Disobedience," the two greatest docu-
ments in the history of human rights, both made a practice of making
phenological observations and keeping phenological journals? Alas,
it probably is coincidence, a correlation unconnected to causation.
But while writing great essays on the rights of humankind, essays that
would help to change the modern world, was a rare phenomenon,
making note of first blooms was quite common. Jefferson, Thoreau,
and countless other men and women did it, usually as part of a broader
personal interaction with the natural world.

Is phenology a science? Philosophers of science, who are self-
appointed to answer such questions, might be tempted to say no. They
might call it, perhaps, a technique. Their friends in sociology of science

might prefer to call it a practice. To truly be a science—in the minds of most philosophers of science—phenologists would have to make use of hypothetico-deductive method in order to make sense of some pocket of nature. "Hypothetico" refers to the notion that there is a potentially profound understanding of the world that can be had by contemplating observations—a hypothesis—and from which novel observations might be predicted, or deduced. Phenology simply hasn't worked this way, although it could have. With sufficient data, more scientists, and a different sense of itself, phenology could well have deduced the fact of climate change in the absence of other scientific disciplines, which could later have confirmed phenologists' suspicions.

Philosophers of science don't get to decide which things count as science and which don't, to their frustration. But scientists themselves do. For going on two centuries now, there has been a broad agreement that physics, supported by mathematics, is top-drawer science if only because it deals with the most fundamental aspects of nature—tiny particles, absurdly fast-moving waves, and monumental forces like gravity and electro-magnetism. The other sciences straggle behind, or so physicists would have it, although biologists working at the molecular level are hot on the heels of physicists. Farther back in the order are chemists and those biologists who study whole organisms or ecological systems as well as geologists and geographers, social scientists, and holding up the rear for more than a century after Morren gave them a name, phenologists.

Those who read the latest literature in phenology or contribute to it, perhaps making use of data from phenological networks, generally identify themselves as scientists of a broader stripe. The boundaries between these sciences are somewhat porous, but their research concerns are often quite different. The largest group of scientists who concern themselves with the study of life is biologists. Biology at present is dominated by science conducted at the molecular level—identifying and understanding individual genes, for instance. But there are other biologists who look more broadly at how species fit into the ecosystems where they are found. Ecologists are interested in this at the most theoretical level. Conservation biologists, a somewhat younger branch of the sciences, have more practical concerns, as do those who work in forestry science and agricultural science. Geographers also ponder phenological questions in subdisciplines like biogeography as well as climatology.

The meaning of "phenology" varies from a simple attentiveness to the seasonal rhythms of the biological world, a kind of grace wherever one may find it, to the rigor of today's integrative science. A fine example of the former is the work of Edith Holden, an illustrator who at the turn of the last century made phenological illustrations as part of her nature notes and illustrations, published as *The Nature Notes of an Edwardian Lady* (1905) and *The Country Diary of an Edwardian Lady* (1906). She made the notes and paintings in the West Midlands region of Great Britain, northwest of London and on the border with Wales. For a page illustrating February, Holden paired a poem ("Mountain Gorses") by Elizabeth Barrett Browning, transcribed in Holden's Arts and Crafts style script, with a painting of common gorse (*Ulex europaeus*), a flowering plant with small yellow flowers. While the value of this page to phenological *science* is negligible, as an example of phenological *attentiveness* it is superb.

Did Edith Holden keep more detailed phenological notes? It's possible that she and many other women living at the time did so. In the United States, Catherine "Kate" Furbish of Brunswick, Maine, developed an interest in botany and devoted her life to collecting, drawing, and painting botanical specimens, most of them representing the flora of the state of Maine, including sketches of five hundred species of mushrooms. Following her death, her art and correspondence were donated to the library at Bowdoin College in Brunswick, Maine, where they are today. It is possible that Furbish's letters contain phenological data, although much of this may only be in passing. Even so, the letters and diaries of women like Holden and Furbish, as well as men whose amateur interests never rose to the level of Marsham or Thoreau, may contain mentions of first appearances of flowers, birds, and so on that could enhance and enlarge phenological knowledge.

When a phenological record is made and kept, whether it was intended for such a purpose or not, it may become useful for new sorts of analysis in time. Ivan Margary, a twentieth-century scientist, pored through the records and published a discussion of the Marsham family's phenological notes and commented on them in the *Quarterly Journal of the Royal Meteorological Society* in 1925. Margary was not looking for evidence of anthropogenic climate change. Instead, he was interested in correlations between weather records and the Marshams' phenological records in a search for periodicity—repeated cycles of temperatures and phenological events, possibly related (in the hopes of the scientists of the time) to sunspot cycles.

It was not the only set of records that scientists looked to in order to a relationship between sunspot cycles other natural phenomena. Phenology has a close cousin in dendrochronology, a technique for dating wood, living and cut, based on the spacing of growth rings. Although scientific interest predated his research, A. E. Douglass, an astronomer at the University of Arizona in the early 1900s, developed dendrochronology broadly in the hope that it would reveal patterns of solar behavior related to sunspots. What the technique showed, however, was a record of climate—wet and dry years seen in the one or more rings that a tree will produce in a year. Archaeologists quickly made use of Douglass's studies to date sites in the American Southwest, where pre-Columbian Puebloan Indians such as the Hohokam and the Hisat'sinam (also known as Anasazi) left timbers in ancient structures.

Archaeologists have used dendrochronology as a tool for dating artifacts, but deeper study provides insights in the climate of the Southwest, which changed radically during the early second millennium of the Common Era. Never amply watered over the past few thousand years, the region experienced very severe droughts in the 1100s and in the 1200s—periods of drought that forced changes in where the Puebloan Indians lived, and in *how* they lived.

Except where records have been kept, such as on the Marsham estate, phenology provides no comparable insight into past climate, and none into paleoclimates. Like the discipline of academic history, it is grounded by a textual record—tens of thousands of observations, recorded with accompanying notes about time and place. But dendrochronology and phenology are similar in that they are systematic sciences. As such, they are useful for providing information about the past—information that, with some ingenuity, might be useful for making predictions about the future.

Still, there is a justification for the way that phenology informs the present.

Early on a day in April 1948, on the same day that he died of a heart attack while working to put out a small fire, Aldo Leopold opened his "shack journal" and wrote that the European white birch, which had been in pollen just a few days before, was spent. The notation follows the word "Phenology," which is underlined, as it was on many pages in the journal—a category. There was another short mention of blood-root that day, and then nothing more. It was the last appearance, a final day's work in the life of a practicing phenologist, although Leopold was much more than that.

Leopold's phenological notes, entered in the careful hand of an inveterate field scientist and over a period of more than a dozen years, number hundreds of pages and thousands of individual observations of birch blossoms and geese, bloodworts and tree frogs, and more. Wisdom gleaned from keeping these notes made up the foundation of the short, seasonal essays at the front of his *Sand County Almanac*, published by Oxford University Press in 1949 and reprinted by Ballantine Books just as the environmental movement heated up in the late 1960s. Painting a word image of January thaw and the problem it poses for a meadow mouse, Leopold borrowed the language of the New Deal and wrote that snow, suddenly absent, "means freedom from want and fear" for the mouse.

Leopold was a man ahead of his time with respect to conservation issues, especially concerning land use, but he was also a man of his time. At occasional points, one reads *A Sand County Almanac* today and winces. Aldo Leopold, *our* Aldo Leopold, smoked cigarettes? Hunted with guns and with bow and arrow? Took his "meat from God"? Spoke inelegantly of the death of a "Chinaman"? Indeed he did. And, without a doubt, he shot a wolf when he was a young forester, watching while a "fierce green fire" was extinguished from her eyes.

At Yale, Aldo Leopold, a midwestern boy, studied forestry. In Arizona and New Mexico he paid his dues, working for the National Forest Service. But the phenologist in him emerged in 1935, after Leopold joined the faculty of the University of Wisconsin, purchased a derelict farm in Sand County, and began to restore it. He, his spouse Estella, and his children were pioneers in ecological restoration, and their efforts were experiments, some of them failed. But in time, the experiment grew into an enterprise. The shack, a more than humble box with a peaked roof and a lean-to, with unpainted vertical board-and-batten siding, and a well with pump in front, is shaded in summer today by mature trees—birches and tall pines—that Leopold and his children planted decades ago.

Taking care to make the distinction, Leopold would list what he had seen in the journals—birds and mammals primarily, counting them when he could and estimating when they were too numerous or disorganized to make a tally. He made notes of temperatures and other weather phenomena. And then he noted first appearances, usually under that underlined category. A little over two decades after his death, his daughter, Nina Leopold Bradley, resumed the records and kept them up until 1991. Nina, along with Leopold's son Carl, joined two other scientists to analyze the data and publish them in the *Proceedings of the National Academy of Sciences*. In the careful language that scientists use, the authors speculated that some of the species recorded over time in the shack journals have adapted to warming, with earlier appearances of changes in range, while others have not. While that may seem like mixed news, some good and some bad, it is anything but. Mismatched phenological responses might well, over time, drive some species to extinction and cause difficulties—known politely as "stress"—for others.

In an essay accompanying publication of some of his phenological notes, Leopold called phenology a "horizontal" science, meaning by that much the same thing that Mark Schwartz does when he calls it an *integrative* science. Phenology cuts across the biological sciences while at the same time demanding inputs from and providing insights to enrich physical sciences such as geology, geography, anthropology, and—yes—even physics.

FIGURE 3.2. The Leopold family's "shack."

Leopold did not, himself, see climate change coming. But he did think about the moral consequences of being a man in a world where most things are not men. In "The Land Ethic," Leopold did not use the language of "dualism" and "antidualism" that I discussed in chapter 1. Instead, he talked about productive land and commodities on the one hand and the land as community, one that included humans, on the other—a biota. "A thing is right when it tends to preserve the integrity, stability, and beauty of the biotic community," Leopold wrote, including himself and us in the community. "It is wrong when it tends other-wise."

The land ethic is, as he rightly called it, "the upshot," the takeaway. But the joy in reading *A Sand County Almanac*, for me, comes in those early pages, the almanac—twenty-two short essays divided unevenly (there are four in April, Eliot's cruelest month, and only one apiece for the wintry months, late spring, August, and September). It was to these essays that the desert botanist Janice Bowers may have been alluding to when she wrote that she recalled wanting to be a naturalist, before she became a scientist, but didn't quite know how at first. She sat on a rock in a stream and "tried to think appropriate thoughts about nature and life," thoughts that eluded her.

Aldo Leopold thought appropriate thoughts about nature and life and inspired a generation of environmentalists and now, perhaps, an-other generation. He had help in the form of charming and accurate illustrations by Charles W. Schwartz, as well as deep insights from making phenological notes, year after year, watching the ebb and flow of solar energy (a phrase he used and understood) sweep through the life of his Sand County farm.

Almost all phenological records through the 1950s—the Marshams', Jefferson's, Thoreau's, Leopold's—provide localized data about cli-mate. For any given data, there are data for one place on a map. Ge-ographers call this point data, and from a geographical perspective, a single point datum is not very interesting. To look for patterns, it's more useful to have many data points spread out over an area—a state, a re-gion, a nation. So while phenology, beginning before there was a word for it, was a longitudinal study—a time sequence measured in years or decades—it was not what scientists call *extensive*. The scale was limited to the perceptions of single observers.

Working from his office at Montana State University in Bozeman, Montana, Joseph Caprio changed that in the 1950s. By training, Caprio was interested in the relationship between climate and weather, on the one hand, and plants, on the other. But Caprio's training wasn't fo-cused in phenology. Before getting a job in Montana, he had worked at

the University of California, Riverside, on weather and citrus production. The relationship between weather and agriculture seems obvious
enough. Citrus growers, for instance, need warning of the possibility of
a frost in order to fire up their smudge pots. But along with other bio-
meteorologists (the specialty in which he earned his PhD from Utah
State), Caprio wanted to provide predictions about yields from weather
data.

Enter the lilac. Common lilacs can be found throughout the Great
Plains and Western states—indeed, across the United States. Once
established, they usually continue to blossom each spring even if many
of the plants with which they were once planted and associated are
long gone. Working with weather observers in the state—a group that
already existed and was trained to make careful observations—and
with garden clubs, Caprio was able to establish a network (the simplest
meaning of the term, by the way—"network" is just a synonym for a
group of people linked by some common interest) of about three hundred people who reported first blossoms, peak bloom, and last bloom,
writing their observations on cards provided by Caprio and mailing
them to him when they were done. From these data, Caprio was able to
make a paper map (there was no geographic information systems technology at the time) of plant climate, or phenology. By the next year, the
network extended to thirty-five hundred observers across the Western
states. With a sequence of maps, divided by altitudes, Caprio was able
to show the average dates that lilacs bloom. In 1966 he published the
findings as "The Pattern of Plant Development in the Western United
States."

Continuing and further extending the network into the Eastern states in the early 1960s, Caprio began sending out clones of two
species of honeysuckle, all started at a nursery in Iowa, for observers to
plant and watch. In time—ironically, just as the small matter of anthropogenic climate change was becoming clear to researchers in fields like
the atmospheric sciences—losses of funding and Caprio's retirement
brought the networks to a seeming close, but they were reorganized
and revived after the turn of this century.

Following his retirement, Caprio worked in Tucson at the dendrochronology lab with the tree rings of ponderosa pine, joining that other technique in order to show that daily weather patterns might be found in tree ring data.

Caprio did more than create networks and expand the scale of phenological research. He continued a long effort to invest phenology with methodological and mathematical rigor. Phenologists, for their part, have suffered from what might be called rigor envy. There are problems with rigor in phenological observations and data, some of which I will discuss in chapter 4, and for which phenologists are presently finding sophisticated work-arounds. But the status of phenology is changing, fast, because of climate change. What was a backwater among the sciences in a world where climate changed with Darwinian slowness is a force in its own right, to be reckoned with in the effort to understand, predict, accommodate, and—one surely hopes—to roll back anthropogenic climate change.

Mark D. Schwartz, a leading phenologist who has spent his career at the University of Wisconsin, Milwaukee, has high hopes for phenology as an "integrative science," one that provides linkages between other sciences. It was Schwartz who revived the lilac and honeysuckle networks that Caprio pioneered, along with researchers for the U.S. Geology Survey and the Scripps Institution of Oceanography at the University of California, San Diego. Motivated in large part by the contribution that phenology can make to understanding anthropogenic climate change, and environmental change more broadly conceived, Schwartz and his colleagues successfully endeavored to breathe new life into the lilac networks and reconfigure them as the USA National Phenology Network (USA-NPN). But Schwartz is serious about his belief that phenology is an "integrative" science, and so the mission of the USA-NPN has been more broadly drawn than its predecessors were. It is a network not merely in the simple sense of a group of people with a shared interest but also in the sense of integrating disparate groups of people and data toward shared ends and yet-to-be-defined interests. For instance, USA-NPN draws data from the North American Bird Phenology Program, which is digitizing—transcribing paper records to digital form—the observations of thousands of birders and other observers in search of patterns, changing and otherwise, among migratory birds in the United States.

These observations accrued over time. In 1881, an ornithologist named Wells W. Cooke asked other observers to provide sightings of migratory birds for his research. Starting with twenty observers, Cooke's project got help from organizations such as the American Ornithologists Union and individuals like C. Hart Merriam (see chap. 5); in time, around three thousand observers were contributing data. Over decades, the number of observers waxed and waned, but the reported observations numbered in the millions—all on paper cards. Paper cards are not useless, but they need to be transcribed in order to suit present-day research protocols. And so, citizen scientists are transcribing them on home computers, one record at a time.

"Citizen science" is a well-meaning but unfortunate phrasing. As a historian of science, I have difficulty using it. I am using it here because, as a matter of usage, there is almost certainly no going back. But as formed, it (*a*) makes an unfortunate distinction and (*b*), even then, doesn't make the distinction that it wishes to make. Let's discuss the latter point first, as this is simply a matter of grammar. And I shall get quickly to the point. Are there people who do science who are not citizens?

The phrase likely traces to the "Science and the Citizen" column in the magazine *Scientific American* and is close in meaning to "citizen arrest." What is meant by the phrase is "an amateur doing science," in other words, amateur science as distinct from professional science. One presumes that "citizen" has been substituted for "amateur" because "citizen" has no pejorative sense, whereas "amateur" can connote sloppiness, dilettantism, lack of rigor, and so on. (The same, unfortunately, might be said of another alternative, "volunteer.") One is advised to consider Charles Darwin "no mere amateur," even though that is—by strict definition—exactly what he was. Even so, the correct construction of this notion would properly be, not "citizen," but "layperson" or "civilian." Lay science or civilian science. The former has perhaps been ruled out because of other meanings of lay, and indeed the word is derived from a synonym of "common." The latter is perhaps distasteful because scientists do not wish to think of themselves as professional militants.

So the construction is backward, unless scientists have decided to deny their citizenship.

The unfortunate distinction is that between the professional scientist and the amateur. This is a rather more complicated business. Think, for instance, about Rachel Carson, who studied the genetics of fishes at Johns Hopkins University and earned a masters degree, but was thereafter never employed as a research scientist (and never earned a doctorate in a field of science). Was Carson a professional scientist or an amateur? Are *The Sea around Us* and *Silent Spring* the works of an amateur, or of a professional scientist?

There is no useful answer to this. It's a foolish question. One could say that Carson was a citizen scientist, but let us call her a trained scientist, one who did important work. And she is far from alone in this. Barry Commoner, a founder of modern ecology, was also a citizen scientist, although he was as much a professional scientist as any.

As I've said, though, this ship has sailed. When discussing opportunities for participating in phenological networks, I use the terms "citizen scientist," "participant," "observer," "reporter," and "volunteer" more or less interchangeably.

Perhaps the best-known program for citizen scientists, one that has had a long existence, is the Christmas Bird Count of the National Audubon Society. Begun over a century ago, in 1900, the Christmas Bird Count was an effort by Frank Chapman, curator for birds at the American Museum of Natural History in New York and a leading conservationist and natural history writer of the time, to publicize bird conservation efforts—the founding of the National Audubon Society would follow in its wake five years later—to tally numbers of birds while at the same time saving the lives of a number of them. The count was Chapman's response to what was then a continuing tradition called the "side hunt" in the United States, in which revelers would go on Christmas bird hunts, teaming up to see which team could bring home the most birds, as well as mammals. In language from a later age, the Christmas Bird Count raised consciousnesses and in time became a holiday season tradition in its own right. The count is organized by local chapters of the society and takes place between mid-December and early January. The data collected provides an annual snapshot of the abundance and distribution of bird species in the United States. Although it is not specifically a phenological event, it is a model for citizen science and a good introduction to organized nature observing for anyone who wishes to participate.

Three large phenological networks for citizen scientists are Project Budburst, Nature's Notebook (part of the USA-NPN), and FrogWatch USA. Project Budburst is a network of observers reporting plant phenology, including leafing, flowering, and fruiting. The organization provides online instruction to help participants learn plants and plant groups. While the group has a list of plants for which they solicit observations, they do not discriminate against invasive or nonnative plants.

In addition, there are many regional phenological networks. I've listed some of them at the end of this book in the section titled "Further Reading."

4

Noting with a Climate Eye

The geese return to Wisconsin in March, wrote Aldo Leopold in *A Sand*
County Almanac. And as long as Leopold's book resides in a trusted
location, on a shelf or tabletop, the geese—Leopold's geese—will continue to return in March. This notation is on page 18 of the edition in
my hands. They return whenever I open to that page, the same way that
Henry goes into the woods to confront "the necessaries" and John Muir
sits and becomes acquainted with a flower for a minute, or a day. These
are experiences recorded by these writers and sustained in print year
after year.

Leopold's geese have returned at some risk, he tells us, in contrast
to the lesser gambles taken by the cardinal and the chipmunk. And
in numbers. How many? The record, after a few weeks, from an April
observation, was 642 geese, counted on the eleventh of the month in
1946. Six hundred and forty-two geese counted on a single day! If that
doesn't excite you, you may have read this far for nothing. I humbly
apologize. But it's likely that, having come all this way with me, you're
savvy to the thrill of counting geese in springtime, or something along
those lines. Or perhaps you want to be.

Learning your plants, birds, trees, frog sounds, insects, and more is
a time consuming, sometimes frustrating, business, sweetened by the
opportunities to spend time with like-minded people (if you especially
like people) or keeping your own company (if solitude is your preference). All the same, it takes time, as do observations and the business
of recording what you've observed. So why do it?

As for any activity that requires time and energy in an otherwise busy
life, there are rewards for getting the work done. Psychologists class re-

wards under two headings: those that count as extrinsic motivations and those that are intrinsic. Extrinsic rewards come from outside oneself. Things like money and honors count as extrinsic rewards. If you engage in citizen science, keeping careful records and reporting them faithfully, you may receive recognition. Other than that, there's not a lot of reward. I have searched the cosmos for an extrinsic motivation for making phenological observations and recording them, and I came up with this: if you live on an estate, and you keep good phenological records (ideally over generations), your records may possibly add to the estate's provenance and increase its value.

Let's move on to intrinsic rewards.

Phenology is exciting! It is, at least, from a certain point of view. Watching for the first leaves to break out in the spring is a little like watching an eclipse of the sun, except that there is far less risk of rendering yourself blind. Watching for the geese to return, whenever that happens, is several things at once. It's an anticipation of the end of something like winter and the arrival of something like spring (no matter where the earth's axis is pointing with respect to the sun). It's a fulfillment of nature's promise to restore and revive. It's a host of honks where honking was absent, and flocks where there was previously only air.

As is reading Leopold, watching for the geese to return is a ritual, and a fine one. Seeing your geese return, or whatever you've chosen to anticipate, is a moment to acknowledge.

Acknowledge it. Write it down.

"Do you keep a journal?" Ralph Waldo Emerson asked of Henry David Thoreau. Henry made it his life's work to answer in the affirmative should the question ever come up again.

Many years later, David Brooks, the *New York Times* columnist who wrote *Bobos in Paradise*, a description and critique of a segment of elite American culture, decided one day to devote his column to people who keep journals, wondering aloud for his audience whether keeping a journal wasn't a sign of narcissism. While he didn't push for the narcissistic interpretation, Brooks did make the argument, with the help of some psychologists, that immediate introspection—who am I? how could I be better?—misleads, and that an unconscious process of knowing oneself is a better one.

This is a very limited notion of what a journal is for, and of why you should keep one, why you *might* keep one. It's a very limited sense, too, of what introspection is. Introspection may or may not reveal one's true character, but at the very least it allows you to devote some time thinking about it. Without something outside your mind, a mirror of sorts, on which to focus your introspection, the distractions of daily life easily take over, and you risk becoming the product of your distractions. David Foster Wallace pointed at this in his justly admired address to the Kenyon College commencement audience in 2005. We all have a choice, Wallace suggested, to see the world entirely from the perspective of *me*, and how every action has an impact on *me*, or to see it as a broad distribution of me-like beings. To practice, in one word, *empathy* (as opposed to just possessing empathy) and to practice it as a matter of choice. This is difficult to do *in the moment*. Reflection, the process of returning to a moment, as when one takes pen in hand and writes in a journal, makes this easier. And is just as much a matter of introspection as anything else is.

Now, Aldo Leopold saw this more broadly, by appealing to a choice: cast your empathy over the whole of the natural world rather than reserving it for humans—or for your neighbors or family alone. To say nothing of reserving it for just yourself, although "just you" doesn't quite count as empathy. Leopold understood that something like a journal was useful as a mirror of the world around us. Here's why: you can try to experience and then forget everything that happens to you in a day—everything you see, everything you do, everything that anyone says to you. There's a kind of nobility in doing this. You can live truly in the moment. The problem is that, barring neurological damage to your brain, you *won't* forget all of it. As we all know, some of it sticks while much of it doesn't, and the archivist in charge of sorting what sticks and what vanishes without a trace doesn't seem to be fully in our employ. With a journal, you get to compete with the judgment of that ornery and uncaring archivist and insist that some things stick. Things like the scarlet tanager you saw on a walk this afternoon or the raccoon you spotted peering out of the storm drain. Those things get to stick around if you journal them and read back (reading back is important).

Without a journal, our remembered lives are like a community theater performance of *Our Town*. There's a stage, a few chairs, and fewer actors than there are characters, some actors playing multiple roles. There aren't any trees or birds, frogs or mosquitoes. Well, maybe we can do without the mosquitoes. But no butterflies, and there aren't really any seasons, either. In New Hampshire! Nothing like the crickets on a warm summer night or the bite of snow blown by a swift wind on a winter day. Just chairs and some people talking.

With a journal, our lives grow so very much richer. There are the ponderosa pines, standing tall, each one appearing to shrug. "Ehh," they say with body language. The two pines you pass everyday that seem to be holding hands. That nuthatch! Is there more than one? It always seems to be the same one. Every time you see it you wonder if nuthatches aren't from Australia, so determined are they to experience the world upside down, from our perspective. You haven't bothered to look it up. The theory doesn't hold water. It's just something silly you thought one day.

Under the pines, a trail, traveled by the oddest assortment of people. There's the fellow who hands out Grateful Dead CDs, and who perpetually remembers you as someone you've never met. The couple who seem so terribly mismatched—she loves the hike and he would rather be at home watching ESPN. The lone woman with the really big dog. Out of the pines, on a patch of grass, a dozen crows stand in a sort of digital pose, equally spaced in a diamond grid. That sound? Chickadee. And that one? Woodpecker. You look around to see the caller—a flicker. You round a corner and see four birds pretending to be the navy's Blue Angels, without the gosh-awful noise. (But you actually liked the noise in the moment, didn't you? Go back in your journal and see.) Across this open patch is a cloud. A hatch of insects! What are they? And clouds overhead, cirrus clouds, high and thin. Weather coming! (Was it just at this moment you thought about the gardening tools you left outside and want to bring in lest it rain? You didn't add this to the journal, so maybe, maybe not.) Back into the pines. As you pass by an old ponderosa, you stop, turn around, walk up to it, get your nose up to the bark, and sniff. Is that vanilla smell present? It's hard to tell today. But the bark is so, what's the word? Rugged? If it were salmon, you'd say it had flaked.

You'll see more, and see more clearly if you record what you see. You'll participate in your life with more agency. So why not keep a journal, if you haven't developed the habit already? And if you do keep a journal, why not expand its scope to record seasonal and phenological events?

Keeping a journal with notes on what goes on in your dooryard is not simply a way of participating in your life, and the lifeworlds around you, on a daily basis. Nature adapting to conditions of acute flux is something new, albeit unwished for. In the past, our parents and their parents looked at nature as an unchanging thing, a solid stage on which they lived out their lives. It has ceased to be that (if it ever truly was). Now, keeping a journal filled with phenological events is a way of tracking a *new* and changing nature, a natural world never before seen or experienced. And while it will arrive in your dooryard at what will often (but not always) feel to you like a snail's pace, the new nature will be coming at light speed relative to evolutionary time.

Would it be best simply to fight climate change and to refuse to acknowledge climatic changes? The answer, in part, and only in part, is no. Climate change is a bad turn of events in human history, but climatic changes to nature are just nature, neither good nor bad but outside human morality altogether. Rage and take action against the former, pay attention to and marvel at the latter. And make no mistake: climatic change will arrive in your dooryard, for every dooryard is a crossroads of the world. On a breezy day, the air you breathe in came from some distant place. The rain that falls evaporated from some body of water in another state, another country. Birds that alight in your trees may have begun their migratory journey across a sea. And not only is your dooryard a crossroads; these natural roadways themselves are also changing. Nature, in its manifold adaptations to changing climate, is presenting you with a new dooryard, day in and day out, and will continue to do so throughout your life, at a rate that modern humans have never before experienced.

Hedgehog cactus (?) buds and bloom

March 26
Taliesin West
Scottsdale, Arizona

In bloom: Hedgehog cactus (I think that's what it is) and brittlebrush. Pencil cholla buds look just about to burst, in a week or two. The flowers of the hedgehog are a deep magenta, emerging from a cone-like bud (there are three buds in this picture).

Saguaro, jumping cholla, and barrel cactuses are waiting their turn. Saguaro won't bloom until summer.

FIGURE 4.1. Page from a phenological journal. While handwritten journals are as beautiful as one's handwriting, and sketches make a useful and attractive visual record, the ability to quickly photograph plants, animals, and landscapes and then import them into a digital journal makes that format ideal, especially if you add scans of your sketches and notes. If you can, print your journal out from time to time, to acid-free paper such as cotton rag. Always make a note of date and place, and weather (especially temperature and cloud cover).

As climatic changes unfold, each and every one of them, it is important that they be noticed, discussed, recorded, compared. Important to whom? Important to everyone, whether they voted Republican or Democratic, pray in evangelical churches, in mosques, or in no church at all. Important to men and to women. Climatic change is important where work goes on and people play, where people raise children and in kitchens and hearths.

Not everyone will pay attention. Not everyone will write things down and keep careful records. Perhaps you will. So let's begin.

Most everyone with an interest in or affection for nature has stopped in their tracks and beheld this world outside ourselves. Perhaps you've sprawled on the grass one summer's day, lazily watching clouds drift across the sky, slowly forming and reforming as they do. Or you've lingered a while, watching ants go about their busywork. Or fish, set on tasks that we don't fully grasp! One day, a friend and I sat on a boulder in the middle of a Western stream, watching tiny fish wriggle through water, oblivious (the fish, that is) to the rarity of that substance in the arid West.

But have you ever looked at a pencil? This is a question that I often ask students in several classes I teach, classes where part of the final grade requires descriptions of things.

Of course my students have *looked* at pencils. But *observing* pencils and describing them using suitable language is an entirely different matter. Pick up a number 2 pencil and describe it. A pencil is long and narrow. Good start. What is the word for the body of the pencil? It's a "shaft," although a "rod" would do. What's the geometric shape of the shaft? (Here you get to put some of that mathematical knowledge you learned in school to good use.)

It's a prism. Specifically, it's a *hexagonal* prism.

What color is the shaft? And what is it made of? What kind of wood? What is the word for that metal thing that holds the eraser in place? (It's a ferrule.)

My students sometimes considered this level of descriptive detail a case of overkill. If you agree, be thankful that you never had Louis Agassiz for a teacher. Agassiz, the Swiss expert on fish who "discovered" the ice ages but opposed Darwinian thought, was a kind of master teacher, if you believe Nathaniel Shaler's story about him. Shaler became a geologist at Harvard University (not the undergraduate college there, but its one-time Lawrence Scientific School); he studied with Agassiz.

Shaler's story goes like this: on the first day of his class, Agassiz placed a pan with a preserved fish in front of each student, with instructions that they discover what they might learn by looking at the fish, without discussing it with other students or doing any reading. After an hour, Shaler thought he had a sufficient understanding of the fish and looked for Agassiz so as to report his newfound wisdom. But Agassiz wouldn't discuss the fish. So Shaler, mildly frustrated, went back to work and continued to look at aspects of the fish each day for a week. At last, ready to report at length on his discoveries—the series of scales, the number of teeth, and so on—Shaler got Agassiz's attention and told his teacher what he had learned.

"That's not right," said Agassiz.

After many more hours of study, Shaler was able to provide Agassiz with an answer that satisfied him enough to allow Shaler to move on to a new task—sorting through bones of different fishes and doing his best to reconstruct several specimens of fish.

Thus did Agassiz, a fine observer in his own right, teach his students to observe and to describe.

What is an observation? How does one go about making one? Must a good observer learn by staring for hours at a dead fish in a pan (as opposed to sitting with a friend on a boulder, watching them swim)? How reliant on observations are the sciences, particularly those that intersect with anthropogenic climate change?

For their sheer world-changing power, there are probably no more significant observations than those made by Galileo Galilei in the years 1609-10 and reported in his book, *Sidereus nuncius* or *The Starry Messenger*. Turning a telescope on the heavens for what were almost certainly the first telescopic observations of the sun, the moon, Jupiter, and Venus, Galileo found that what he saw through the device seemed to be in agreement with the model of the solar system (he didn't call it that, but we do) devised by Nicolai Copernicus and published in *De revolutionibus* more than half a century previous. Today, historians and astronomers agree that Galileo's observations did not prove the Copernican system, but they surely pointed in that direction.

The Copernican system itself was the result of mathematical ideas applied to countless individual observations of the stars, the sun, the planets, and the moon, observations made by many observers over centuries. Not all of them were made in the interest of what we now call astronomy; many were made by would-be astrologers, looking to tell fortunes to the highest bidder. But the observations were there when Copernicus looked to make sense of them.

To be useful, to provide data for the Copernican system and our current understanding of the solar system—where a spacecraft, an object about the size of a Chevrolet Suburban, can be sent hurtling from one end to the other in order to photograph Pluto and not miss the mark—it wasn't necessary for every observation and measurement to be dead accurate. But most of them needed to be. Science is built on a foundation of care (in Latin, *curare*, the root of the words "accurate" and "curator") that every observer can, and many observers do, bring to bear on the natural world.

Galileo made observations, whereas Copernicus relied on books filled with the observations made by others before him. Charles Darwin made many observations while circumnavigating South America, as well as on the grounds of his home in Down, but he also relied on observations that others made—often at his request. It is possible to do great science without making many observations firsthand. But it is not possible to do science without recourse to *someone's* observations. Even theoretical physics is conducted with the real, observable world in mind. And so it is with phenological observations.

A phenological observation, when it is useful to science, consists of four parts. First is a correct identification of a species of plant or animal. The second is a note on the calendar day of the year, a consequence of being in the right place at the right time—the moment a plant blooms, a bud bursts and begins to form a leaf, a bird that has been missing for a season or two appears in your dooryard or a nearby pond or marsh. The third (not always, but often) is description, such as the color of a leaf. The fourth is what scientists call metadata: a report with the correct notation of *time of day* and *place*, without which the report is useless to science.

Before getting specific about individual observations, let's look first at how data is used by science, working backward from a conclusion. I promise: what follows is about the leaves of birch, ash, and cherry trees and currant shrubs. It will take a page or two to get there.

In the year 2000, two German scientists, Frank-M. Chmielewski and Thomas Rötzer, analyzed phenological data collected from across Europe and wrote a paper in which they concluded that warming of one degree Celsius (1.8 degrees Fahrenheit) in mean annual temperature will lengthen the growing season in Europe by five days. If that warming happens during the early spring, between February and April in Europe, the growing season lengthens by *eight* days. The "mean annual temperature" generally means pretty much the same thing as average temperature, but here it is significant. Annual mean temperature can increase by one degree in a number of ways—perhaps mostly in a particular season and less so in other seasons. This may happen because of seasonal circulations—patterns of weather events that recur each year. In the United States, for instance, the monsoon in southern Arizona, which brings rainfall to the Sonoran Desert in July and August each year, is an example of seasonal circulation. For Europe, the North Atlantic Oscillation is a key circulation that determines both Europe's weather and its climate.

Chmielewski and Rötzer also concluded that, in many regions in Europe, the growing season has increased by eight days over the past three decades. This means, of course, that early springtime temperatures have increased over the same period of time by more than one degree Celsius (or close to two degrees Fahrenheit).

Having so concluded, the scientists put their pencils down. They don't say whether a longer growing season is a good thing or a bad thing. They don't say whether eight days is better than five days, or worse than five days. Drs. Chmielekski and Rötzer might tell you what they think, if you ask them, but as phenologists drawing conclusions from data, they believe the upshot to be outside their official purview. So we are mostly free to draw our own conclusions from their conclusions.

Something else goes unsaid in their conclusion, although they could well have said it. The finding—growing season lengthens by eight days when early spring temperatures increase one degree Celsius—is the result of a sort of natural experiment. The "sort of" is important, because laboratory experiments can be repeated and must be repeatable for experimental findings to have validity. In the case of conclusions from phenological data, we can't easily order up a second planet identical to Earth, age it 4.6 billion years, and then warm it up a tad by increasing its greenhouse gases. Even so, the changes to which the planet are being subjected are *like* an experiment, and the conclusions that scientists draw from such data tell us about not only warming but much else besides. They tell us how ecological systems work and whether the systems are static (their physical characteristics and mix of species remaining intact) over specified periods of time or dynamic.

Ecologists have devised experiments of various kinds, but never at the scale of global warming. Robert MacArthur, along with Edward O. Wilson, experimented with patterns of ecological succession on an island; the results gave him his theory of island biogeography. Mike Gilpin, at the University of California, San Diego, experimented with fruit flies in laboratory habitats. Anthropogenic climate change, for better or for worse (and I'll show my hand here—it's for worse), is the biggest experiment anyone has devised, if anyone could be said to have devised it. And it's accidental—unplanned and unintended.

Let us work backward from the conclusions. Before concluding that they had demonstrated a relationship between temperature and growing season over the past thirty years, Chmielewski and Rötzer declined to make predictions based on their analysis. They pointed out, guardedly and with an unspoken tip of the hat to Heraclitus, that not enough is known about forest growth to simply project their findings into the future.

But they did take care to pay attention to what part of their finding might be due to a straightforward increase in air temperature in any

particular place, and what part to changes in the North Atlantic Oscillation, Europe's weather maker in spring and fall, which seems to be
bringing springtime to Europe about 3.5 days early each decade.

The bulk of the scientists' article considered these variables: temperatures, natural regions, and data from International Phenological
Gardens. Okay, one variable at a time. The temperatures came from an
archive of climate data kept by the National Center for Atmospheric
Research. Europe's natural regions are like the growing zones in the
United States but organized on the basis of similar soils, climate, and
vegetation, rather than on climate alone.

The International Phenological Gardens are in a smattering of locations throughout Europe, a network like the USA-NPN mentioned in
chapter 3, but with a much longer history. Begun in the 1950s, they
consist (at this writing) of eighty-nine gardens in nineteen countries.
A small variety of genetically similar plants grow in the gardens—
twenty-one species at present, including *Betula pubescens* (silver birch),
Prunus avium (wild cherry), *Sorbus aucuparia* (mountain ash), and *Ribes
alpinum* (a currant). Observers report on eight different phenological
stages, including the beginning of leaf unfolding, beginning of flowering, general flowering, first ripe fruits, autumn coloring, and leaf fall.
Between beginning of leaf unfolding and beginning of flowering are
two stages that are not self-explanatory.

For their study, the scientists Chmielewski and Rötzer selected data
collected by the IPGs specifically related to the moments that leaves of
the three species of tree and a shrub—the aforementioned European
birch, wild cherry, mountain ash, and currant—opened. In order to
have those data, someone needed to be present at each garden to observe and record the date each opened.

And so that is the prize. A leaf opens and you make a note of it. Your
note and many thousands of others are aggregated as data (the plural of
datum, that is, your note) and, when analyzed with other data, lead to
the conclusion that spring comes eight days earlier when there is a rise
of one degree Celsius in average temperatures early in spring.

Now, to be useful to *you*, all that is required is to be present and to
pay attention. You do not need to engage in citizen science in order to

take pleasure in, and edification from, phenological observation. It's enough to know that you are seeing the world around you, that you are aware of your world and, in time, of the way it is changing. But as luck would have it, an observation that's useful for science requires you to be present, to visit your site every day or two, and to pay attention.

There are some differences between the kinds of observations one makes for science, and other kinds of observations, and that's what this chapter is about. For science, one must ideally make objective observations, without bias, and record them with care.

When observing outside of science, there's more than a little room for fantasy, for emotion, and for creative, even magical, thinking. In a sense, Henry David Thoreau engaged in this kind of observing related to his acknowledged status as an American transcendentalist, although he did less of this as he matured as an observer. And John Muir observed both within science and without, never (or seldom) confusing the one kind of observation with the other.

When thinking about how science works, I often think of the relative values of the Richter scale and of the Mercalli scale. Each in its own way provides a sense of "how big" an earthquake was, whether it's a shaker that I feel under my feet or one that I hear or read about in the media. The Richter scale is a mathematical technique for quantifying the energy released in an earthquake. Richter measures ignore the consequence that the earthquake may or may not have had on people. The Mercalli scale (more correctly, the Modified Mercalli Intensity scale) grades the effect that an earthquake has on people and their environs. One might think that news media would report Mercalli and ignore Richter. But one would be wrong. The chances are that you have heard of Richter and not of Mercalli.[1] Even if you have heard of Mercalli, it is rare that a Mercalli number is part of any news story.[2] What you tend to get instead is a Richter magnitude and a verbal report such as "no

---

[1] There is a close analogy between my discussion of Richter versus Mercalli here, on the one hand, and the thrust of my discussion of "global" climate warming and dooryard phenology, on the other. Science, in its effort to get to the truth, has to average variations, in place and in time. This is good for getting at the truth, but not always for understanding truth's portent.

[2] The U.S. Geological Survey "publishes" Modified Mercalli Intensities on its website under the title "Did You Feel It?"

injuries or damage to structures was reported." That's about a V, on a scale of I-XII on the Mercalli scale, or weaker.

It's the difference between the two scales that's important. Richter is a logarithmic calculus based on wave amplitudes, as measured in a seismograph. The logarithm itself is reason enough to leave it out of media reports. A magnitude 8.0 earthquake isn't one unit larger than a magnitude 7.0 quake (to take just one intuitive reading of the difference). It's a hundred times more energy. A magnitude 7.0 releases a thousand times more energy than a 5.0. More important, however, is that there is no direct correlation between magnitude and the thing we care about: what damage did it do? A magnitude 6.1 can do a lot of damage if there is a population area nearby—considerably more than a magnitude 7.1 would do if it occurred in a remote, mountainous area. And there are plenty of other variables in terms of impact. Wet sand is much more effective in transmitting shaking to structure than solid bedrock is. A mountain range between the focus of an earthquake and a population area will absorb a lot of energy. The Richter scale doesn't care about these differences, and that's exactly why it is useful for studying earthquakes.

Modified Mercalli Intensities (MMIs), in contrast, are the reports of eye (and body) witnesses—citizen scientists, you might say, but the U.S. Geologic Survey asks only "did you feel it?" The MMIs reported on the U.S. Geologic Survey site are effectively crowdsourced. They do little to help seismologists in their effort to someday predict earthquakes, but they provide the information that an interested party wants. Did you feel it? Yes, some chimneys were broken and toppled. That's an MMI VII.

We would be much better served if the media reported earthquakes in Mercalli, and if we knew what the scale measures, leaving Richter magnitudes to scientists who are doing science.

Do phenologists have the equivalent of seismographs? In fact, they do, in two senses. Remote sensing, using photographs from satellites, provides some phenological data, such as the progress of green-up. But the observations of phenological events made by citizen scientists serve a similar function and provide (*a*) data that remote sensing cannot or (*b*) data that keep remote sensing honest (the latter are called "ground truth").

As was the case for seismology, there is a difference between what scientists hope to take away from phenological observations and what *we* would like to know about our dooryards. Scientists are hoping to build ever more rigorous and accurate models of how ecological units respond to changing climate, as well as hoping to better understand the ecological units themselves. The rest of us want those things as well. But, just as we care more about whether there was damage from an earthquake and therefore are better served by reports of Modified Mercalli Intensities, so, too, are we somewhat more alert to our own observations that, for instance, there seem to be fewer deer (or more deer) passing through our dooryards, not just this year but for the past three years.

Whether your aim is to participate in the growth of scientific knowledge or to know your dooryard, what should you observe, and how should you record it? If your aim is to participate in citizen science alone, the various networks will provide you with *protocols* for observing and for reporting data. Most make this as simple as possible to do but still look for close attention to detail. Project BudBurst, for instance, asks you to identify an individual tree, in the case of deciduous trees, by common name and by Latin name, and to provide a location in longitude and latitude as well as city and state. From here, you can provide what the project's researchers refer to as a "single" report of a phenological event. The form asks: "What is your plant doing now?" And you answer by checking a box, perhaps: "Many leaves have unfolded from the buds." And then you turn that in. Nature's Notebook is similarly organized to simply reporting. Single reports might be a way to start out in citizen science or to scratch the itch to do a little phenology if you're traveling for business or pleasure.

"Regular" reporting is rather more elaborate. The regular reporting form asks for information about the tree, such as whether it is a hundred feet or fewer from a building or from concrete or asphalt, whether the tree is regularly watered (irrigated), what kind of habitat it is growing in (a park, natural setting, school, your backyard), and what the shading

	A	B	C	D
1	Date	Temperature	Observation	
2				
3	20-Jul-15	59 degrees F	Bobcat! When I woke up there was a bobcat right outside	
4			the cabin window. I think it was watching rabbits in the yard.	
5			Very peaceful. It took a long time to notice I was looking at it.	
6			Finally noticed I was watching, so it slowly got up and walked	
7			away, in no hurry whatever. I was able to take some really bad	
8			photographs. The camera focused on the screen instead of	
9			the cat. Emma sat on the other window sill, looking the other	
10			way, and never even noticed the cat.	
11				
12				

FIGURE 4.2. A spreadsheet, such as Microsoft Excel, is a durable means for keeping careful records. It is searchable, and with enough categories it will nudge you to provide important data like dates, weather, and temperatures. A spreadsheet isn't especially friendly, but it can be the staccato to a handwritten record's legato.

is like. Then the form asks that you record, for this tree, the month and day of each of the following phenological moments: bud burst, first leaf, all leaves unfolded, first flower, full flower (the moment when at least half of the flowers are either open or releasing pollen on three or more branches), the first ripe fruit, full fruiting (half or more of the branches have fully ripe fruit), 50 percent color (half of the leaves have changed), and 50 percent leaf fall.

Project BudBurst has similar reporting forms for wildflowers and herbs, conifers, evergreens, and grasses. All told, the network currently has a list of two hundred fifty plants from which to choose.

A thorough report on the regular form requires a degree of commitment. Project BudBurst does not demand that you fill in all of the blanks in order to file a regular report, but let us assume that you want to do so, or at least as complete as you can. And since you're going to the trouble, why not observe multiple trees—individuals of the same species, spread out along your phenological trail, as well as multiple species.

With this sort of commitment in mind, it's best to turn to a spreadsheet program like Microsoft Excel, naming individual trees on your trail, one row per tree, in the order you encounter them, and columns for each of the phenological events. Your spreadsheet might look like the one shown here. From this, you can fill out regular reports as you please, for as many plants as you care to.

Networks for reporting birds, frogs, and other animals have similar but, in some cases, stricter protocols. FrogWatch USA, for instance, ask that volunteers register an observing site (this is similar to the BudBurst questions about location) and include information about the weather conditions while observing, then begin making observations only after remaining quiet for at least two minutes. When you are ready to begin, the protocols are precise. You cup your hands around your ears and listen for exactly three minutes, remaining still the entire time. (It's obviously a good idea to make yourself as comfortable as you can be before beginning.) At the end of three minutes, you record the time you began listening, the time you stopped listening, the species you believe you can correctly identify, and the call intensity for each of the species. FrogWatch USA also asks that if there is a disturbance (from a loud airplane or a cellphone ringing) that you start the observing procedure over again. The network takes observations using forms that you fill out online.

The requirements for reporting to the FrogWatch USA network are not onerous, but there may be days when you hear frogs, or expect to, but don't wish to follow all the guidelines. In such cases, note the calls in your journal. Similarly, BudBurst doesn't ask for reports on mushrooms (which are not plants) or mosses, so if these organisms interest you, report them to yourself in a spreadsheet or your journal, doing so with as much metadata as you can. Someday, someone might be interested.

When you begin making phenological observations and recording them, your attention and understanding will evolve. It makes no difference whether you report your observations as a citizen scientist or keep them to yourself, in your journal or even in a spreadsheet. You will improve. If you keep records primarily for yourself, a system might be of some use. Many ecologists, conservation biologists, and others agree that Joseph Grinnell, director of the Museum of Vertebrate Zoology at the University of California (Berkeley), devised the gold standard for keeping nature notes.

Grinnell's system has three parts: a journal (no different in kind from the journals discussed in this chapter), a catalog, and species accounts. The catalog refers to the metadata associated with specimens that

Grinnell and his field associates collected. If you engage in collecting of any kind, of insects for instance, or leaves, then you will want to keep a catalog, although it won't be extensive. More likely, you will keep lists, as birders do: a life list, an annual list, and lists of birds counted on, say, a Christmas Bird Count. Like these, your lists will include information about the locations of your observations.

Finally, Grinnell insisted on species accounts: detailed descriptions of observations of individual organisms in their natural settings. There are some species accounts in the chapters that follow, but they are guides only. You will become familiar with species, learn from more detailed written accounts, and add your own observations to them, observations that may range from the rigorous to the fanciful.

At first, the process may feel a bit like doing a school project. But with passing time—months, years, decades—monitoring your dooryard can change into a cornerstone of who you are, of where you've been and where you're going. And that's just looking inward. At the very same time, your observations and notations will stand as testament to how the lifeworld around you is transforming. Not all of the latter changes will be caused by climate. But many of them will be.

5

Bedrock and Baselines

The coming of snowfall: on a day before a full moon, the sky draws close. The ceiling drops, darkening to the color of tarnished silver. By four in the afternoon there are pioneers. Soon they are followed by settlers, which look a little out of place, even lost at first, but not for long. In these first hours, it is just snowfall. By evening it is a storm, gaining strength by the hour. Families count off: two, three, four. Safely home. Overnight, the winds pick up and fall back, like ocean swell. By morning the first act is over, but the drama does not abate. At an hour after sunrise the inside of the house is filled with light. If you are lucky, other than to shovel you will stay in.

Through that day, the sun licks the snow. Not for nothing do chefs and bakers speak of icing, frosting, and glazing. And then, at sunset, a full moon rises on the opposite horizon. This is what we have been waiting for—the spectacle of reflected light. The glazed snow glitters and shines. The world, in just that moment, that splendid moment, is renewed.

Change. Landscapes change in just this way and through hundreds of other events, some cataclysmic and some in ways just like this, changing for a few short hours or days, and others for keeps, or seemingly so. Some changes are good from a certain perspective, maybe from yours. Others are not as good. We live for some, while dreading, occasionally fearing, others. The right mood, the perfect time of day, can make a rain shower into a miracle, while too much rain and too little preparation can bring disaster.

It is never just the weather, of course. It is what the weather does with the plants and animals, the microbes, the soil, the underlying rock.

Count the contingencies: winter or spring, summer or fall. A forest of trees and understory. Or fields after the harvest. Wet, heavy snow or flakes so delicate they nearly refuse to fall at all. Temperature, relative humidity, wind speed, barometric pressure. Sand, soil, or stone. Cliffsides and gentle slopes. The list of contingencies goes on. It is customary to think of landscapes as reliable, rooted in solid ground and (while moody perhaps) unchanging in ways that reassure. Absent an earthquake or some other catastrophe, our dooryards feel permanent enough. But landscapes never are this, although some will change more rapidly than others. In any case, we tend to take landscapes for granted, rarely devoting much time to make a list of each contingency, each this-depends-on-that. These are the elements of change. They, not permanence, make our dooryards what they are.

What of climatic changes, then? How will the changes in climate transform your dooryard and mine? How deep, how lasting will the changes be? Will each change be but tiny, each difference adding to those before? Answers to these queries can be predicted in some gross and abstract ways; but together with other instances of it-depends-on-that, we can only watch, wait, and see.

As we watch and wait, it is reasonable to ponder: what is the nature of change itself? Seasons change and landscapes change, but are these changes *real*? Our planet has warmed before and has cooled down again. Are these actual changes? Or are they something different? Is change real, or are changes just cycles, like seasons—part of something *larger* than any single change, something unchanging in itself? This is an old question, one of the oldest in philosophy. Go back more than two millennia and you will find two philosophers staking claims to each side of it. Heraclitus and Parmenides (who are known as pre-Socratic philosophers) defined the terms of a debate that Parmenides, for the most part, won.

Heraclitus, for his part, held that all was flux, that change was the only constant. You can only step into the same river once, he was quoted as saying, both metaphorically and not. Step into it a second time and it has changed. It is not the same river. Parmenides demurred: change, he said, is an illusion. And while that makes little sense intuitively (change

certainly seems real enough when it happens), Socrates and his aco-
lyte Plato came down on the side of Parmenides. Change is illusory,
is always underlain by the unchanging, the permanent, the foundation. The universe is not a river but a sturdy house. Philosophy and its
daughter, modern science, agree on this point. Change, at some level,
is simply an expression of some underlying law, some unyielding and
permanent process.

Phenology seems a case in point in Parmenides's favor. The seasons bring change. Trees bud, buds become leaves. Branches once bar-
ren are now filled with growth. Birds build nests, lay eggs, raise chicks.
Chicks fledge and fly away. As summer becomes autumn, the leaves
turn color and fall. Where it is cold enough, the rains become snow. It
all repeats anew the following year. Change, and yet unchanging.

Does this mean that climatic changes, the manifold consequences
of climate change, have no reality? Here is an answer: one thing that
paleontology shows is that the chance that any given species will go
extinct is 100 percent. Yes, someday polar bears will become extinct—
along with every other species. It is simply a matter of time. This is the
unchanging thing, but it is of little consolation to the polar bear, or to
any other species member who departs with the rest of its clan because
natural selection has zeroed it out in the Anthropocene. In this sense,
Heraclitus was right: the change we experience is real and ever present, even if it is the consequence of universal laws that are themselves
constant through time. Rivers change from moment to moment. They
swell from the spring melt, flood whole towns, and sometimes take
lives.

The reality of change is a matter of perspective, of where one stands.
Figure 5.1 compares rates of change of the basic features of any given
dooryard: bedrocks, surfaces, ground and surface water, soils, plants,
animals, weather. It simplifies matters quite a bit, as do all such graphs.
In this one, change happens more rapidly the higher one goes on the
vertical axis, less so the deeper one plows. Our best assurance that
change is real is that we can measure it and compare one change with
another. This chapter provides an overview of these components of the
landscape and provides suggestions about ways that you can create a

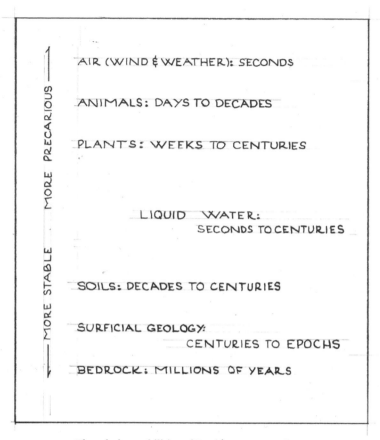

FIGURE 5.1. The relative stabilities of Earth's component layers.

baseline for noting and measuring change over the coming years and decades.

Bedrock. Bedrock changes, but it does not change quickly. (If you happen to live in the path of lava flow from an active volcano, as in Hawaii, take the previous sentence as an obvious overgeneralization.) Geologists have classified bedrocks by periods, which are measurable in time. Beneath my dooryard in Maine, the bedrock was Ordovician in age, meaning that it has been around (although not at the surface) for close to 500 million years. The bedrock underlying the cinders at my Arizona dooryard is Permian—perhaps 275 million years old. You can discover the age of the bedrock at the foundation of your dooryard and the period in which it falls, if you wish, by searching the Internet (enter "bedrock," "geology," and the name of your town in the search engine) or by purchasing a map from your state geological survey. Geological maps are not difficult to read, and are invariably colorful. Because bedrock is the most permanent aspect of your dooryard, it is perhaps the least interesting part of your baseline. But it is interesting to know all the same. In many places, fossils from past geological times are not far underfoot.

The shape of the land changes more quickly. In northern states across the United States, the landscape has been shaped by ice that retreated in some places a mere ten thousand years before the present. Glaciated regions are covered with drumlins, eskers, kettle lakes and ponds, moraines of various kinds, and other products of the most recent ice age. Learning the meanings of these terms and identifying the features that correspond to them enriches your understanding of the surrounding landscape. Residents of eastern Washington state may know that parts of the landscape there, known as the channeled scablands, formed in an enormous flood, with a headwall perhaps five hundred feet high, that occurred when an ice dam breached, emptying a lake in a very short time. Knowledgeable coastal Californians look at the mountains around their dooryards and associate them with the earthquakes they experience from time to time. Other coastal dwellers, such as those who live near the beaches in North Carolina and South Carolina, have seen islands form and disappear as a result of hurricanes. Who can forget the devastation that occurred in New Orleans in the wake of Hurricane Katrina?

The study of the surface of the earth is called, not surprisingly, surficial geology, and the branch of geology that studies those processes that make the surface what it is, is called geomorphology. Depending on what place you call your dooryard, there is a chance that it will go through some geomorphological change over the decades to come. Increased precipitation (a generalized expectation in a warming climate) may lead to erosion (although it might just as likely provide denser ground cover, preventing erosion).

Just as there are geological maps of bedrock, so, too, there are maps of surficial geology. These, too, are colorful but perhaps not as intuitive to read. If you belong to a garden club, or an environmental organization with local chapters, it is well worth the cost of a small stipend to engage a local geologist for a field trip. Although geomorphology today is largely quantitative in practice, an older tradition told stories about landscapes—stories rooted in time, structure, and process. Geomorphologists, even those of the quantitative sort, have fascinating stories to tell. Rivers, beaches, lakes, hillsides, shorelines, and other features of the earth's surface tend to be, in a geomorphologist's mind, a sort of motion picture in which the present is a momentary freeze frame. There are few places in the United States that have not changed in profound ways over the past few thousand years.

Soil, the varied top layer of "solid" ground, changes even more quickly than does the surficial geology. Gardeners commonly think of soil as the product of their own toil as they turn it, adding various amendments and compost, working to achieve a proper acidity for the plants they wish to grow and harvest. But soils will change without the helping hand of a gardener in a *succession* from purely physical sediments to a biota in their own right, teaming with microbes, animal species, and plants. In time, the primary succession from sediments to soil will be disturbed by fire, by the introduction of new species (with or without an anthropogenic cause), and by climatic change. It is the last of these that interests us. Climatic changes may bring about higher soil temperatures at new times of the year, causing a *secondary succession* in the soil, that is, a change from one kind of soil to another, with a revised set of microbes, plants, and animals.

Liquid water in bedrock and on the surface of the land is highly variable. Vernal pools, some as large as ponds or lakes, appear in springtime, as their name implies, and disappear for the rest of the year, along with much of the biota they briefly support. Ground water moves, but large aquifers may impound water for millennia. Lakes are ephemeral on a geological timescale and may change more quickly during droughts and periods of heavy precipitation. Terry Tempest Williams's *Refuge*, a natural history of the Great Salt Lake region in counterpoint to the story of her mother's cancer, uses changes in the level and shoreline of the lake (which is itself merely a remnant of the far larger pluvial Lake Bonneville) as a marker of the passing of time.

Rooted in the soil, plants change with rapidity, and animals even more so. Air, on a windy day, scarcely rests for a second, but that is the subject of a subsequent chapter. We can think of plants and animals, en masse, using two concepts in ecology: life zones and succession.

Life zones. Almost anyone with a respectable interest in biology has made a pilgrimage to the Galapagos Islands, venturing forth in their imaginations if not in actual fact. One half imagines that, while gathering together for one reason or another, virtual pilgrims, together with those who have managed to make the actual trek, raise their glasses and toast: next year in the Galapagos. This is a fine sentiment, and one that I share—to make a journey to the very place where, according to the myth that he promulgated in his autobiography, Charles Darwin got the idea for natural selection by observing variations among the finches that populate these islands. Even those who know that this was not the case, that the road to a theory of evolution was more complex, that Darwin was oblivious to the significance of the finches until long after he returned to England, still celebrate the Galapagos. There is value to myth, and I respect it. Equally important, the Galapagos are a cluster of contingent landscapes that drive evolution. By the time he wrote his autobiography, this notion about the significance of the Galapagos had fully crystalized in Darwin's mind.

It seems a shame, then, that Little Spring in Arizona is not so highly regarded. I have successfully made the latter pilgrimage once, and I attempted it more than that. Little Spring is to ecology what the Galapagos are to evolutionary biology; at least I like to think that this is so. On my first pilgrimage there, I missed it entirely. But I was thoroughly entranced by the journey, via a National Forest Service fire road, through Hart Prairie at the western base of the San Francisco Peaks north of Flagstaff, Arizona. The aspens were golden in autumn and stark against the blue sky. The next time, a full decade later, I was more successful just because I did something I wouldn't do the first time around. I asked a forester for directions.

"Go up this way," he said, pointing down a dirt road carpeted in fallen aspen leaves. "You will come to an opening. When you get there you'll see a two-track. Follow that up into a stand of Douglas fir. You'll find a plaque."

I didn't quite know what a two-track was, but I could guess. As I reached the open meadow, I found a promising double trail, tire tracks: a two-track. I followed it into the stand of Douglas fir and found the plaque without much trouble. I laughed when I read it. The text on the plaque was more of a tribute to the people who placed it there than to the historical circumstance it commemorated. Two dates are given along with an abundance of text, none of which have anything to do with the significance of the site. No matter. I had found it. This was the site of C. Hart Merriam's base camp in the summer of 1889, the place where the life zone concept (and its progeny, the biome) was born. This, when he and his assistants were not out collecting, was Merriam's temporary dooryard. How much had changed since that summer more than a century before, I could not say. I felt certain that the stand of aspens on the far side of the meadow weren't there and would not have presented the golden spectacle of autumn leaves that I was witness to. But other aspects might have been similar—not the same trees, but the same species. The camp was in a stand of Douglas fir in 1889, according to Merriam's description, and they were in a stand of firs when I saw them.

In 1889, the U.S. Department of Agriculture dipped into discretionary funds and found $600 to support an expedition designed by Merriam to survey the biological wealth of the American West. For their money, Department of Agriculture got one of the great bargains in the history of science. It was too little money to do more than make a drop in the bucket with respect to mapping the biological resources of the West, much less the United States. But Merriam emerged from the summer expedition with something else—a comprehensive theory of the distribution of species that we know today as the life zones theory.

With his wife and with assistance from Vernon Bailey, Leonhard Stejneger, and locals hired from the incipient town of Flagstaff, Merriam spent the summer traversing the San Francisco Peaks and the surrounding Colorado Plateau, from deep inside the Grand Canyon across

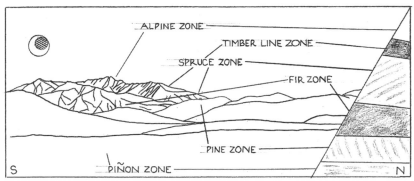

FIGURE 5.2. C. Hart Merriam's original life zones were based on an expedition in the southern Colorado Plateau. Merriam described the zones at higher altitudes by plants and animals he collected in the San Francisco Peaks, shown above as seen from the east. Each of these zones hosts a different climate, resulting in a unique mixture of plants and animals. With few exceptions, populations of organisms will have to migrate to higher altitudes or northward in the United States in order to remain within the climate to which they are adapted.

the plateau to the top of Mount Humphreys, and into the Painted Desert to the east. He applied methods that the great German naturalist, Alexander von Humboldt, developed in South America; Merriam recorded measurements of temperature and barometric pressure, slope, and temperature and noted the relationships between these and the organisms he found. From those relationships, over the course of the summer of 1889, Merriam generalized seven distinct life zones.

In his conception of the life zone, Merriam had done what pioneers in science do: he had devised a first approximation of order from seeming chaos—that is, he originated a very good idea. It was much too orderly, but it was a good idea, and it provides a road map to climatic change. As a rule, as mean global temperature increases, zones will move northward (if north of the equator) and southward (if south of the equator). And they will move upslope in either case. Species that can move with their zones will survive; those that cannot will become extinct.

The U.S. Department of Agriculture adopted Merriam's system of zoning to develop their map of plant hardiness zones for North America, which gardeners know well, just as they know that full sun, partial shade, and full shade make a difference in what to plant. A glance at the department's map shows what's at stake: the boundaries

FIGURE 5.3. Almost all gardeners, and all farmers, are familiar with a map like this one, showing the eleven plant hardiness zones in the United States, according to an understanding of temperature, altitude, and latitude not much different from those that C. Hart Merriam worked out after his summer at Little Spring, Arizona. The U.S. Department of Agriculture provides the map and revises it periodically to keep up with climatic changes. The map above is not as detailed as the department's maps and provides no more than a sense of the zoning across the continent. To do more would quickly date this book, as these zones are changing. I have withheld zone labels for the same reason.

of each zone are sinuous and convoluted. As climates warm, these twists will not simply move north. They will change according to local conditions, just as they were initially inscribed on maps with reference to local conditions. Assuming that change is kept somewhat in check by more enlightened policy making than is evident at the time I write this, gardeners (and farmers) will experience only the mild inconvenience of having to choose different seeds, different varieties, forgoing some vegetables in exchange for adopting new plantings.

Wild plants will be at more of a disadvantage. As wild plants reproduce, they do so in what seems to many of us as an inefficient process. Saguaro cactuses may each produce as many as a hundred thousand seeds, only one of which may survive into adulthood. The survivors, in their early stages, are known to foresters and botanists as recruits. Climatic change will reconfigure the conditions in which recruits survive. Generally this will be upslope or northward (in North America), all else being equal. But slopes are never infinite; they top out. And landscapes are fragmented. Plants, in the process of migrating to compensate for climatic change, run up against four-lane interstates, farms, suburbs, and cities. When they do, they run out of luck.

The add-in is contingency, the state of "it depends." Landscapes are contingent; they are contingent on a great many factors. The simple idea of changes in altitude in Merriam's very good idea is expressed in nature in slopes, which twist and turn every which way. Think of a ravine or deep valley running from east to west. The north-facing slope will be shaded in parts for much of the day, changing the application of the simple rule. Shade has an effect not just on photosynthesis but also on soil temperature, in varying degrees. A rising annual mean temperature may change the soil temperature but not the exposure of plants to sunlight.

Moreover, the migration of plants usually occurs over centuries, not decades (although more rapid changes do occur, and I will discuss this later in the chapter). There are already initiatives afoot to help nature through managed migrations, but the task is daunting and is currently unpopular with some climate change activists.

Succession. Ecologists (and others) use the term "succession" to refer to a series of changes in the makeups of soils, plants, and animals in the landscape. Some of these changes are regular, and some are contingent. Both Parmenides and Heraclitus would draw interesting conclusions from a study of the successions caused by beavers and the transformations they bring to the landscapes of North America, now and in the past. Heraclitus would point to changes, day to day and year to year, that beavers initiate in the landscape. Parmenides would counter with cycles, and the longer-term unchanging nature of things. Both might ponder the introduction of trappers, which decimated populations of beavers for a time and brought the cycles and the short-term changes to an end. You may have encountered this story as a child, but it's worthwhile repicturing the full process, because it is the essence of cyclical landscape succession.

Picture in your mind a forest in a modest valley, wide and not too deep. There may be some openings in the tree canopy, but this is clearly a forest, possibly made up of maples, oaks, and birch. Trees and the wildlife they support dominate the landscape. In the midst of the forest, a stream bubbles and flows as it makes its way downslope. There may be fish—minnows or larger fish—in the stream. There are crawdads. Moss grows on rocks in the stream. Were you to step in it, it might well be cold.

One day, a beaver discovers the forest, makes a note of the stream, and surveys the abundant resources. The next day, she returns with other beavers, who begin to gnaw through younger trees with their teeth and then drag the trees into place across the stream. Before long, they have created the framework for a dam across the stream. They begin to infill the frameworks with smaller branches and mud. As the stream begins to back up behind the newly engineered dam, the beavers construct lodges in the area that will soon enough become flooded; placing the openings to the lodge below what will become the waterline. Soon enough, the lodges are complete and water sufficient to create a pond has impounded behind the dam.

Out in the pond, some trees that were too large for cutting remain standing; in time, they will die, but in the meantime they make ideal aeries for predatory birds, hawks and eagles, who feast as a result of the revised ecology. Where there were woodland species with a riparian strip (a grouping of plants and animals that only live close to water, in this case, the stream), there are now a host of new aquatic plants, invertebrates, and vertebrates. Insects provide meals for dozens of species of birds that would have bypassed this place before the beavers came.

It is a pocket paradise, but with each passing year, spring floods make work for the beavers, who must maintain the dam in good working order. With each spring flood comes sediment, which falls to the floor of the pond in the low-energy environment. Each year, this raises the bottom of the pond. The dam must be built higher, and wider perhaps, to do the same work that the original dam did. The lodges, too, must be adjusted.

Then, one spring, a fifty- or hundred-year flood washes out the dam. Some of the beavers do what they can to restore it, but others scout nearby woodlands and find a suitable spot for creating a new beaver pond. Before long, the old pond is abandoned. The dam, in disrepair, ceases to hold back the flow of the stream. The floor of the pond, now exposed, changes from mud to soil. Seeds, some of them carried by animals and others wind-borne, settle into the soil and germinate. Before long, the pond floor has become a meadow, filled with flowers and grasses in spring and summer. The meadow is a popular browse, where deer sample the plants. With such abundance, the deer can afford to pass over the woodier species of plants in favor of tasty shoots. A pack of wolves settles nearby, enjoying the deer. All of the animals make use of the stream, which has reestablished itself in a somewhat different course.

Small pockets in the pond floor still fill with water in springtime and host frogs into the warmth of summer. The sky is busy with birds, some nesting there, others in the old forest that ringed the beaver pond. In time, the woody plants that the deer pass over become saplings, and the saplings become trees. The trees pass through a succession of their own until finally this is once again woodland, with a few openings in the canopy but otherwise, a forest. And one day a beaver visits. Parmenides is pleased. Nothing changes. Heraclitus shakes his head. Everything has changed.

The cycles of succession that beavers engineer are dramatic, but every landscape is the consequence of succession, sometimes in progress (where a disturbance has occurred) but often in a condition that ecologists loosely call climax. Whether climax landscapes will undergo a new process of succession in response to climate change, and how those successions will unfold in case they do, is an open question—and one of the reasons that phenology as citizen science is gaining standing.

History and archaeology. Fires and beavers are not alone in bringing change to the landscape. Humans are engineers, too, and have been transforming the North American landscape for around fourteen thousand years.

When I lived in Maine there were occasions when I took my household of cats—Phoebe, Grebe, Oscar, Charlie, and Cloud—for short hikes through the woodland behind my home. They seemed to enjoy this, even though the activity required frequent coaxing on my part. Cats have an innate sense that, really, this is just too far from home. They would protest. I would encourage. A few words were enough to keep them moving forward. Invariably, we would come across stone walls in the woods, monuments to industry in an otherwise natural setting. The cats intuitively treated these as clear barriers to further progress.

At first, the walls made no sense to me. But stone walls are part of the mythology of Maine woodlands, and in no time I knew about their provenance. At each wall I encountered, I was witness to four processes. The first was glaciation and the retreat of ice from Maine around ten thousand years ago. The ice left behind a rubble field extending for tens of thousands of acres. In an older tradition, geologists referred to the rubble as "drift." Later, it became known as moraine. Next, forest succession changed the landscape into woodland. Then, in the eighteenth and nineteenth centuries, farmers cut the forests here, partly to clear the land for pasturage and partly to provide wood for construction and for heat through the long winter months. It was because of their interest in pasturage that farmers moved the morainal rocks, using stone sleds in the winter months and constructing a variety of stone structures usually no more than three feet tall.

Finally, agriculture moved west. Pastures, no longer needed, were left fallow. Forest succession renewed. New England was transformed once again into a vast secondary forest—with odd stone walls separating patches of woodland with little difference between the patches.

On the other side of the continent, in the desert landscapes of California, one may encounter, while hiking, piles of stone with a quite different explanation. These piles are ancient fish traps, created by Native Americans thousands of years ago to impound fish in the vast, shallow lakes that formed here during the last ice age and that disappeared completely only a few hundred years ago. (Some such lakes, such as Lake Owens, persisted into historical time, only to disappear into the pipes and aqueducts that feed the farms and urban environments of California.) Coastal areas in many parts of the United States have archaeological remains of technologies developed by Native Americans to impound fish, but to find such devices in the clear absence of water in the present day is a little unsettling.

Local history and archaeology are useful resources for developing an overall picture in your mind of your dooryard's past. While the differences between history and archaeology aren't important for holding that picture in your mind, they are quite different for the people who practice them academically and professionally. Historians, to an overwhelming extent, ground their work in textual evidence—letters, notebooks, journals and diaries, newspapers, court records, and books. Anything that is written and preserved is grist for the historian's mill. Archaeologists, in contrast, look to physical evidence as the basis for making claims about the past. Tools, middens (trash piles), and other kinds of physical evidence are the raw materials of archaeological knowledge. In some cases, as in projects involving historic preservation, historians and archaeologists set aside their differences and collaborate toward common ends. But much of the time, they work in the same places using different intellectual tools. It is often the work of some third party—a museum curator, a ranger, a naturalist—that brings their work together.

One tool that archaeologists use would leave historians scratching their heads. That tool is pollen. There is a whole subdiscipline of science devoted to the study and classification of pollens—palynology— and a sub-subdiscipline that focuses on ancient pollens. The latter is called paleopalynology. Paleopalynologists collect pollens from the layers in which they are preserved at the bottom of a lake or pond and, from them, develop a picture of the set of plants that grew and in what abundance at some time in the past. Archaeologists tend to use paleopalynology as a tool for correlating sites (that is, showing the relation in time between two or more sites) and for painting a picture of the contemporary ecological landscape for a site. But paleopalynologists are interesting in their own right. If you ever have an opportunity to talk to one, take advantage of it. The chances are very good that a paleopalynologist can reconstruct, in considerable detail, a vision of your dooryard or someplace nearby, as it was some time in the past. Was it once a meadow, or woodland? A paleopalynologist can provide the answer.

Large parts of the American landscape went through a fairly cataclysmic transformation in the few centuries following European contact. Enterprising individuals, making use of their own labor or the hard work of free men and slaves alike moved rivers, shaved mountains, replaced whole ecosystems, drained lakes and wetlands, and created roadways to move goods and resources from one end of the continent to another. The local impacts of the centuries of industry interest us here, as do the changes wrought by Native Americans before European contact. Currently, the resources available in print for the lay reader are somewhat patchy, but to get an idea of the *extent* of change, the exemplar is William Cronon's *Changes in the Land,* an overview of the transformation of the New England countryside.

Nothing comforts the soul so much as knowing where to begin. For the complete phenologist, the point of departure is best marked by establishing a baseline.

The next few pages, like the bulk of this chapter, are not about phenology, per se, but they are helpful both for making phenological observations and for appreciating their meaning, not simply over time but from month to month. The idea is to take Heraclitus and Parmenides out in the field with you to create an inventory of the changing and the unchanging, the precarious and the stable. Heraclitus would have you see that all of it is precarious, while Parmenides would have you focus on the stable, warning that you should not miss the forest for the trees. (Yes, he will talk in clichés just like that. He invented some of them.) As field assistants, they are a mixed bag. Neither is at all useful for carrying gear or even lunch, and they both tend to be biased and overbearing. But, in general, they will aid your observations and notes more than they will detract.

While it is not necessary to create a baseline in order to record phenological change over time, or to be mindful of the present time, it is a useful early step to take, and not just for recording change. It is a way to, as progressive agronomist Wes Jackson has put it, become native to this place, if one has not already established that valuable sense of rootedness. In this chapter, we look at the geological history of our dooryards, the recent geographical circumstances that create these places as they are today, and the life zones in which they are placed— as well as the changes in the physical, biotic, and human landscapes that may come about over future decades. We will then look at three aspects of creating a baseline: mapping your greater dooryard; establishing one or more transects that connect your dooryard to the greater landscape; and using repeat photograph, or rephotography, to record change visually, establishing the archaeological and historical roots of the place where you live.

It is a good idea to establish a baseline now. In words Heraclitus might have used, it's time to step into the river once and take note of it before stepping into the river again, and again, in order to measure change and to learn from this unfolding experiment. These dooryards, real and figurative, yours and mine, are set within larger landscapes about which many things are known. Poets and essayists may have written about them—think of Thoreau, Muir, and Abbey—and scientists have certainly measured and described them, collected them and shrunk them down to the size of lists and maps. But the devil, it is said, is in the details, and the details are the essence of this book.

Repeat photography. Beginning around the turn of the last century, farms and rangeland in the American Southwest began to experience a process that locals called "arroyo cutting" (*arroyo*, which sounds pleasing, is Spanish for *ditch*). Streams and washes would suddenly erode and deepen, taking vast acreages of arable soil with them. Was this a consequence of grazing, or of overgrazing? Was something else at work? Geologists, such as Kirk Bryan, from Yale University, attempted to answer these questions, but the variables were complex. The climate of the Southwest has changed over the past 150 years, only partly due to anthropogenic factors, and this must be taken into account along with land use. With no clear solution in sight, a historian at the University of Arizona named James Rodney Hastings began a research project to find an answer. Hastings found photographs from the 1870s, 1880s, and so on. With some ingenuity, he was able to take photographs at the same positions and with the same angle of view as those he had found, giving him a comparison photograph. Most of Hastings's "repeat photographs," published with coauthor Raymond M. Turner in *The Changing Mile: An Ecological Study of Vegetation Change with Time in the Lower Mile of an Arid and Semiarid Region,* are so accurately matched that one can scan the earlier and later photos and show them in sequence in a PowerPoint presentation to dramatic effect.

As a way of showing dramatic change, rephotography is better suited to the open landscapes of the American West than to other parts of the United States, but repeat photographs can be helpful for noting change anywhere. To create a baseline, simply create a marker using a permanent stake or, better, a six-inch-square bit of concrete with the center clearly marked. Place a camera with a known focal length on a tripod over the marker, using an improvised plumb bob (a nut or another heavy object tied to one end of a piece of string) to center the camera over the marker. Be sure to make note of the height of the camera above the ground or marker. With a compass, take photographs in sufficient number to complete a panorama, making notes of the compass angles for each shot. Then repeat this process on a regular basis, as often as you wish, but at least once a month for the first year. Thereafter, repeat again on the same days of each year.

If you already have photographs of your dooryard, you can try to match up the camera position, focal length, and date in order to establish a baseline earlier than the present. This may require some experimentation but will deepen your repeat photographic record. Picture Post, part of the Digital Earth Watch network, is an opportunity for citizen scientists to make and submit repeat photographs.

A map or aerial photograph. A detailed, scalable map of your immediate environs is a helpful thing. Such a map can be drawn as a sketch map, to scale or not. But aerial photographic views, made in two or more seasons, are an improvement and are, if you are so inclined, quite a bit of fun to make. The obvious way to create an aerial view of your dooryard is to use a drone; the technology is currently developed to make this a relatively simple option, albeit an expensive one if you must purchase a drone with camera. But rentals are possible. For instance, as of 2015, Photojojo, in San Francisco, was renting a drone with digital camera for under $100. The downside, for those who do not live close to the city, is that you must take a lesson in flying the device before you can rent it.

A lower-tech option, and one that may be a bit more fun, is kite aerial photography, a hobby that has developed with the Internet. The guru of this type of photography is Charles Benton, professor of architecture and the University of California at Berkeley. Benton has been taking kite aerial photographs for more than a decade (at this writing) and posting them to his website, along with many webpages devoted to the equipment and techniques required to make quality images of this sort. There are also web retailers that specialize in a full range of equipment. Images of your dooryard, made at different times of year using kite aerial photography, can be repeated (although exact repeats are difficult) for comparison over time.

A phenological trail. The most useful task you can undertake in advance of keeping phenological records is to establish a phenological trail—a line through the landscape that extends from your dooryard, which is a point on the line. It need not be a straight line. Zigzags are fine if they pass through varied landscapes and biomes. The best way to make sure of that variation is to change altitude, even if this involves only a few feet. The close-in part of your phenological trail should be a distance that you can traverse every day. Farther out, you might want to cover ground once or twice a week or, farther still, at least once a month. As you establish your phenological trail, get to know all of the plants along the route. Ignore the distinction between naturally occurring or native plants, on the one hand, and garden or farm plants, on the other. All are grist for the phenological mill. The camera is, once again, your friend. Digital photos cost nothing, and you can, if you wish, photograph every foot of your phenological trail.

Depending on how much of the foregoing work you take on, you will have made a substantial investment in knowing your dooryard. Photos, maps, and learning are an outlay of time and effort, even if they are inexpensive in dollars. What becomes of that investment if you move, say, in five years? Or in twenty? If you have found your phenological work worthwhile, you will no doubt begin again in your new home. Perhaps you can persuade a neighbor, or the new occupant of your former dooryard, to continue your records and to share them with you. And you can

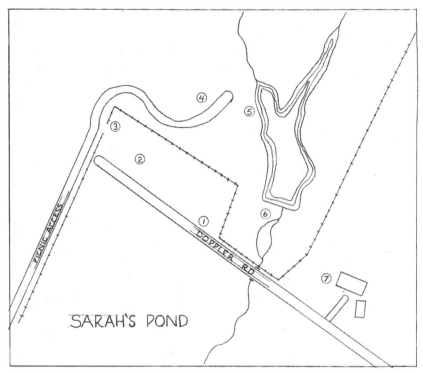

FIGURE 5.4. Sketch map of a hypothetical phenological trail at "Sarah's Pond." It is not a blazed trail but, rather, a set of points along a gentle, repeatable walk. The first point (*1*) is a parking place. From there, walk through the cemetery (*2*) to the gap in the fence (*3*). Note lilacs along the fence line. Pass over into government land and through the trees into the picnic grounds (*4*). If the grounds have been used recently by picnickers, watch for raccoons and scavenging birds. Continue out to Sarah's Pond (*5*), where there are usually egrets and herons and a variety of ducks. Look to see if the osprey nest on the opposite shore is occupied. Walk through open forest along the shoreline and down the bank to the small pond (*6*). Listen for redwing blackbirds and frogs. Retrace steps, but before getting into car, walk past Ferguson's farm (*7*) to see what is blooming in the orchard. Say hello to old Ferguson if he's out in his dooryard.

contribute them to the USA-NPN. The simplest answer is to say that, by developing a baseline and phenological records, you have indeed become native to this place. If you move, any return, on any interval, will be a homecoming of sorts. And should you remain, you'll be witness to all changes, great and small, the fortunate and the unfortunate.

PART | 2

6

The Green World

A hedge of common lilac purrs in springtime. Some might say it buzzes,
but I say otherwise. I discovered this in a roundabout way. Once, when I
had appointed myself as stepfather to seven cats—two white, two white
with smudged foreheads, one black, one gray, and one that was deaf,
with long white hair—my feline brood became a hostel for an infes-
tation of fleas. In order that they (the cats) not long suffer, and favoring the cats over the fleas, I made the cats suffer briefly, bathing them
each using flea soap and placing each one in turn outside in sunshine.
Then I bombed my apartment with some potent proprietary insecti-
cide. And I sat with the cats out in our dooryard as they dried, and the
three hours required for the flea bombs passed by. We played, all of us,
next to a hedge of lilacs, ten feet tall at least, and I remember that the
hedge purred loudly—the sound of countless unseen bees going about
their spring business. I hadn't noticed this before, the purring or the
bees. Perhaps it was all timed to occur together—the cats, the fleas, the
bees, and the springtime blooms of common lilacs.

Are there lilacs in your dooryard? Along your phenological trail?
Planted by homeowners and homesteaders across the United States for
practical value, for color, and for fragrance, common lilacs (*Syringa vul-
garis*) arrived here in the holds of ships from Europe, where they originated in Eastern Europe and Asia. In the colonies, and the states that
temporally followed, lilacs made colorful hedgerows, natural fences,
and foundation plantings wherever soil temperatures dropped predictably below freezing. Jefferson mentions in his Garden Book that he
planted them, and George Washington made a record of transplanting
lilacs in his garden.

They conquered the West—the lilacs, that is—venturing across the prairies, over mountains, and to the Pacific Coast, first to purchased land and later to homesteads and would-be homesteads. Would-be landowners, taking advantage of the Homestead Act of 1862, often planted common lilacs as part of their larger effort to improve what they hoped would be *their* 160 acres. Many homesteaders abandoned these properties, before or after they'd taken their deeds at the end of five years, leaving the lilacs to pass through many years and phases untended. Others persevered in places that still have residents today.

Today, lilacs are a leading research tool for phenological networks. Other plants aid the effort. Climate is best seen in the set of plants that associate with one another in a landscape, and although changes in these associations have other causes not related to climate change, climate change can nevertheless be noticed in the new appearances of plant species, the disappearances of others, and the general health of plant species across the board. This will be more noticeable in some places than in others, and climatic change is only one cause among several. But the plants in your dooryard and your surrounding environment complete your baseline understanding of where you live.

You should know, for instance, your biome and the local plant associations where you live. The U.S. Forest Service is a good resource for identifying your biome and has a four-part hierarchy. *Domains* are areas of the country with similarities of climate, temperature, and precipitation, in particular. *Divisions* are further differentiated by precipitation and temperature. *Provinces* are divisions based on having similar vegetation. *Sections* are the division at the finest scale, hinging on terrain features. But some researchers prefer to speak of ecoregions rather than biomes.

This hierarchy says nothing about the size of the land area of each component. Provinces may cover very large land areas or very small ones, as may sections. In addition to situating your dooryard within a larger region, these divisions are another reminder that climate and vegetation are closely linked and semistable with respect to one another in ways that weather and, for that matter, many observations of wildlife are not.

With that as a start, begin to identify and watch your plants. The

FIGURE 6.1. Wheel of phenophases. To use it as a way of marking the current phenophase, copy to a sheet of heavy paper (card stock), cut out, and put a thumbtack through the center dot. Follow the phenophases for a single plant by turning it so that the current phenophase is at the top of the circle.

great beauty of plants, from a citizen phenologists' point of view, is that they hold still for a little while, unlike warblers and weasels. You can get up close to them and chart their progress through a year's time. You can start out getting to know a handful of them this way, then add a few more each year until you've charted the whole orchestra. As you do this, you'll develop an appreciation for the understory, the part of the forest you may not have seen for the trees. You will see flowers in bloom through the eyes of their pollinators. And then, over time, you will track the changes taking place.

Every plant in your dooryard has a phenology, just as every plant has a life cycle. While it is tempting to look at your phenological trail as a jumble, composed of plants run amok, try thinking of their individual phenologies as the parts for a specific instrument in a symphonic score. The ferns are flutes; the low-flowering shrubs are clarinets. Dogwoods are second violins, maples are trumpets, oaks are bass drums, and, well, you see where this goes. From green-up to senescence, each individual part gives color to all the other parts, coming in at different times, resting for a few measures, mixing, blending, never quite balancing. There is always some dissonance. And you, the phenologist? You get to read the whole score, if you care to. Or just take it in for a few measures. That's good, too.

If you know a bit about plants generally, in addition to knowing a few individual species, much of this chapter will strike you as altogether too elementary. If your knowledge is rudimentary, you may be shocked to discover that botany may be the most linguistically challenging of all of the biological sciences, not simply for its Latinate phylogenetic nomenclature (and there's a mouthful right there) but across the board. Dipping deeply into the study of plants demands a vocabulary that a comparable study of, say, birds doesn't require.

The nomenclature of botany is abstruse, so much so that the proper botanist might take the trouble to point out that I don't really mean nomenclature, I mean terminology. A basic, entry-level textbook explains the workings of plants using terms like "allopolyploidy," "aneuploid," "anisogamy," "apomixis," "archegonium," and "autotetraploidy." And those are just the *A*s. Something like this is true of all of the sciences. For instance, although such terms are fewer in number in physics, to understand physics one really needs to know calculus. I mention this vexing lingo only because I plan to work around these terms wherever I can. I call this making the science accessible. Your college biology professor might call it "dumbing it down." I wouldn't even mention it if botany weren't so overgrown with terminological kudzu.

And if that weren't enough, there is also the problem of fuzziness. Many of the terms that scientists use mean something pretty close to the same thing that other terms signify. One example: "biome" and "life zone" are pretty much the same thing. It's difficult to tell where "shrubs" stop and "trees" begin, although it's pretty clear that a rose bush is a *shrub* and a live oak is a *tree*. Those of us who are not trained as scientists sometimes expect scientists to live up to higher standards, to attach clear and precise meanings to words so that we can tell things apart. But they don't. You have to learn the language, just as you need to learn that one scissors is scissors and two scissors are scissors.

A lot can be learned from reading, but spending time in the company of *one who knows* is worth the time and effort.

Some words: a "tree" is a plant with woody trunk and branches that grow from it. A "shrub" is also woody but is often shorter in height than a tree and seems to be mostly stems, with no clear trunk. All the remaining seed-bearing plants ("angiosperms") have herbaceous stems and branches, meaning that they are *not* woody or are succulents (cactuses). Herbaceous angiosperms are either "forbs" or "grasses." Except: not all grasses are grasses. The grasses ("graminoids") include grass, sedges, and rushes. But let's back up to "forbs." Most wildflowers, such as Indian paintbrush and sunflowers, are forbs. It's a handy word.

The key distinction in botany, for the purposes of phenology, is the difference between angiosperms and everything else. Angiosperms are flowering plants. They are abundant in American landscapes and provide plenty of opportunity for making observations of phenological events. "Everything else" includes conifers, ferns, and even things that aren't plants but appear to be, such as mushrooms.

You can go about learning the plants in you dooryard a number of ways. You may already know several of them, but may I suggest that you check even these, the plants you believe you know well? Many years ago, shortly after I was married, my significant other and I went hiking through the low mountains in Acadia National Park off the coast of Maine. After a couple of hours of hiking, we stopped for a bit of lunch. As we gazed out over the landscape, I noted that there were a surprising number of yellow jackets nearby. "Those are bumble bees," my significant other told me. No, I insisted. I had called them yellow jackets all my life. They are so called, I pointed out, because they appear to be wearing fuzzy yellow and black jackets. You know where this is going, don't you? Certain that I was right, I made a wager. Later, in town, on a visit to the library, I learned that my certain knowledge of stinging insects wasn't certain in the least.

One way to learn plants is to spend time with one or more people who know their plants. Make sure that their knowledge of plants is more assured than my knowledge of bees was. In some locales, there may be published lists, with pictures, of common local plants. There are also guidebooks, some of which arrange plants by the colors of their flowers. However, there is no better way to identify plants than by using a *dichotomous key*. A dichotomous key explores dichotomies that pop up in nature. A plant has seeds, or it doesn't: two choices. If it has seeds, the seeds may be arranged in cones or on stems. If arranged in cones, the trees they come from may be conical, or they may be more palm-like. And so on. Using a dichotomous key, you can guide yourself to a correct identification, which you can then confirm using photographs or other visual materials.

While you learn, then, you may see most or all the plants in your dooryard. But don't count on it.

It is well known, especially among botanists but also to environmentalists more broadly, that modern people—those of us who live in cities and suburbs, who buy our food in grocery stores and dine in restaurants—are virtually "plant blind." Show someone a photograph of a bear walking through a meadow, surrounded by dozens of forbs and trees, and ask what that person sees. "A bear." That's the most likely answer. It doesn't occur to most people that the bear is but one of dozens on dozens of organisms in the photograph. As it happens, it's not simply that we give greater attention to the bear (or some other animal) than to the plants surrounding it. Much more, we don't see many of the plants at all.

I discovered my own plant blindness in a clear and measured way. Some time ago, I paid a visit to Taliesin West, Frank Lloyd Wright's winter encampment in Scottsdale, Arizona. It was there that I began noticing and paying attention to desert plants, beginning with the most obvious (because it was both bizarre and at the same time a kind of icon of the Wild West), the saguaro cactus. I was there for the architecture, of course, but that initial visit led me to read widely about Wright and his apprentices. Several accounts of life at Taliesin West mentioned plants, such as the one for which Wright named the camp he built and occupied in 1928-29 while working on plans for San Marcos in the desert, a project that collapsed with the stock market crash. Wright called the camp Ocatilla, his mispronunciation of the ocotillo, a key Sonoran Desert plant. In one passage, I read about the many yellow flowers of the brittlebush surrounding the Taliesin drafting room in spring. When I read this, I wondered whether the flowers were still growing there. So I opened the shoebox containing my photos (digital photography was just on the horizon) and indeed, brittlebush blooms surrounded the drafting room. I hadn't noticed them. I was blind to brittlebush, and to many plants.

In part, plant blindness is an example of what the twentieth-century philosopher Ludwig Wittgenstein calls "aspect blindness." There are many distinctions between things—aspects—that we don't notice until someone points them out and names them. A cultural example shows how common this is. Walk the streets near the center of most American cities and you're likely to pass homes and buildings that are Gothic in style, or Greek Revival, Craftsman, Federal style, or Romanesque. There are many such distinctions between architectural styles, but they aren't obvious until someone walks us through the varying aspects. Once we see the differences, it's often surprising how they stand out. Wittgenstein called this "aspect dawning." It dawns on us that Gothic looks nothing like Greek Revival.

And so it is with plants. But botanists, alert to the depth of the blindness they find in nonbotanists, blindness to plants in particular, have looked further into the matter. James Wandersee and Elisabeth Schlusser published a short editorial with the title "Preventing Plant Blindness" in the journal *American Biology Teacher* in 1998 and followed up with a longer discussion of plant blindness in *Plant Science Bulletin* a couple of years later. Wandersee and Schlusser started not from Wittgenstein but from their own experiences—and, one imagines, frustrations—as teachers, reflecting ruefully on the mismatch between the importance of plants, on the one hand, and the lack of interest that their students show for plants, on the other. What causes this, they wondered? Why aren't people who are "afflicted" with plant blindness more sensitive to aesthetic qualities of plants, some of which are, according to their list, "adaptations, coevolution, colors, dispersal, diversity, growth habits, scents, sizes, sounds, strength, symmetry, tactility, tastes, and textures"? The botanists found their answer in problems of perception, in the complex relationship between the eye, the brain, and memory. Some of the causes of plant blindness, they claim, are specific to plants. The surfaces of plants, for instance, reflect light in peculiar ways that vary from plant to plant. But other reasons for plant blindness have to do with how little of what we perceive is actually stored in memory. Wandersee and Schlusser quote a Harvard psychologist Stephen Kosslyn who wrote that "the mind is not a camera, the brain is not a VCR."

Simply knowing that we are prone to plant blindness is a major step toward seeing plants. But a camera is a good observing tool for working with plants in your field of view. A photographic record of your entire phenological trail is helpful. In time, you may be able to identify every plant. But photographs of individual plants are especially useful (and in many cases better that bringing specimens home). To make good photographs, it is sometimes helpful to use backgrounds of white or sometimes black paper or poster board so that leaves, buds, and blossoms stand out clearly.

ANGIOSPERMS

Flowering plants, the angiosperms, can be divided into two groups: annuals and perennials. Annuals start from a seed, germinate, grow to maturity, reproduce, then die. Perennials remain alive after reproducing, have a period of dormancy, and green-up again in a new year. Perennials are themselves of two types, woody plants and herbaceous. Woody plants have hard stems that survive aboveground through senescence. Herbaceous plants remain alive underground but will die back aboveground.

Trees, of course, are among the woody perennials, as are a variety of shrubs. In both cases, woody plants may have individual root systems, or their woody parts aboveground may share a root system below. In *The Oldest Living Things in the World*, Rachel Sussman describes Pando, a single quaking aspen with forty-seven thousand stems, looking to the untrained eye like a forest of forty-seven thousand trees. A colony of clones (a clonal colony), Pando has a shared root system that spreads out over 106 acres in Utah. Other woody plants that may look like individuals but can be clonal include black locust, lowbush blueberry, creosote bush, forsythia, hazelnut, honey locust, poplar, sassafras, and sumac.

The period of photosynthesis in deciduous trees has no surprises. Cool-temperate leaf-bearing trees begin photosynthesis in winter, increase the activity rapidly in spring, continue at a maximum through summer, then diminish rapidly in autumn until trees have shed all their foliage. In desert regions, the paloverde, with its evergreen bark, continues photosynthesis through all seasons. Over this period, trees go through a secondary growth, adding a ring to their girth.

Buds are pledges. They promise future leaves, flowers, and primary growth in woody plants. Because of these assurances, they have a special place in the hearts of phenologists because, from buds, all new life from a plant issues forth. Lilacs, for instance, may have a flower bud surrounded by leaf buds at the end of a shoot and along the length of older growth. New buds form in summer and are dormant with the plant over winter. The low temperatures of winter, however, get the buds working, and in springtime they swell to bursting. When they burst, the green-up begins. Bud bursts tend to follow an order, each species of tree or shrub greening up in its own time. In northern New England, it's forsythia that usually signal the start of a new season of growth.

Over the course of a year or so, you will establish the order of bud bursts in your dooryard, as well as the variation within a species and patterns of bud bursts, occurring at different times for different species. It's an exciting time of year, made more so by being attentive to the order of events.

The buds of temperate-zone deciduous trees remain dormant through the winter until temperatures have been cold enough for long

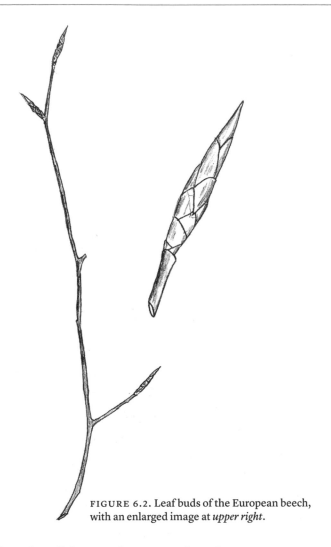

FIGURE 6.2. Leaf buds of the European beech,
with an enlarged image at *upper right*.

enough and until days are long enough and temperatures warm. The
cold enough/long enough part of this cycle varies from tree species to
tree species. Phenologists call it the chilling requirement. If you think
about it, there is really no better example of natural selection at work
than this. Among all the individuals of a species, there are likely to be
some whose buds burst at suboptimal chilling requirements. Ordi-
narily, these individuals may not reach maturity or produce seeds. In
a changing climate, however, some suboptimal chilling requirements
may prove optimal, and these trees may reach maturity and reproduce,

while the existing forest of mature trees may never again experience temperatures that meet the chilling requirement. Is this what will happen over the coming century? It remains to be seen. This is part of the unwitting experiment.

Come autumn, photosynthesis declines and the leaves of deciduous trees lose their green color, which comes from chlorophyll. Chlorophyll isn't stable in leaves over time. The plant must synthesize new chlorophyll in order to maintain a green life and benefit from photosynthesis. As day lengths decline and temperatures fall, the plant stops maintaining chlorophyll. The green-pigmented molecules decrease in the leaf, revealing other pigments (the dominant color among deciduous trees is yellow). Changing pigments in the leaves of some tree species (sugar maples, for instance) follow a complex path. Others (such as deciduous oaks) go straight to brown. The leaf comes to its living end when the plant seals off the point of attachment with a leaf scar, and the leaf falls to the ground where, if it remains, it decomposes.

But I'm getting ahead of myself. Flowers develop in plants as a response to day length (or length of night), also known as photoperiod, or to temperatures. Most spring flowering perennials in temperate part of the globe will flower in response to rising temperatures. Common lilacs initiate flowering from a temperature cue; this is one reason they have such an important place in phenological study. As you begin your studies and record keeping, it's a good idea to watch the development of a flower from start to finish, from budburst to fruit, making notes about the timing and appearance of each stage in the process. Time-lapse photography is quite revealing but provides a false sense of the time involved.

Flowers have such a common place in modern life, as short-lived decorations in homes and places of business, for longer displays in cultured landscapes, and on gowns and lapels of humans in the (generally) nonrecurring phenophase known as "prom," that it's quite easy to forget that these symbols of natural beauty are organs for biological reproduction and are offered as such in a physical metaphor.

If you have never done so, or if you did so at some distracted moment ever so long ago, there is great value in taking the time to dis-

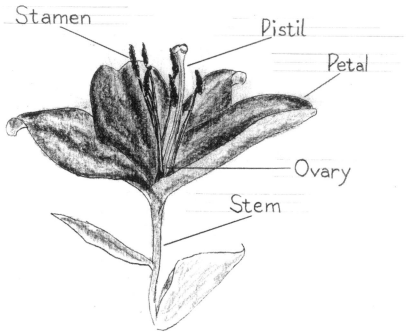

FIGURE 6.3. The reproductive parts of a flower. The stamen is made up of an anther (at the tip) and a filament. The pistil consists of a stigma (at the tip) and a style. The ovary, with ovules within, is also part of the pistil.

sect a flower and to become familiar with its reproductive parts. You'll need a very sharp knife, a magnifying glass, and an undisturbed half hour. After quieting the mind, start by smelling the flower, looking at the arrangement of the sepals, the green leaflike parts that surround the outside of the petals. Count them, and look at the arrangement of the whorl. Next, the petals themselves. Diagram them, if you would like to create a record of your dissection. Once more, count and observe the arrangement. How many petals are there? How do they overlap? Peel one away and look at it in isolation. Feel its texture, examine any veining you can see and contemplate the colors. Then peel the remaining petals, fully revealing the stamens and the pistil.

Stamens first: these delicate organs are made up of male anthers and aptly named filaments. The anthers are the pollen producers. Look at one, zooming in with your magnifying glass. Count the stamens. Feel the pollen between your fingers. (Never discount the worth of feeling things between your fingers. It's harder to forget a texture than to re-member many visual aspects of shape.) Now, remove the stamens to reveal the pistil fully.

It's time to get surgical. With the female pistil on a cutting surface of some kind, slice down the center of the pistil to reveal a cross-section of the ovary, with the ovule inside it; the stalklike style; and the stigma. It is here that pollen, deposited on the stigma by wind or breeze, a bumblebee, a bat, a hummingbird, or any number of pollinators, starts its journey through a pollen tube in the style to the ovule, where it fer-tilizes an egg to form a seed. In self-pollinating plants, such as peas, the pollen can come from within the flower or from an adjacent plant. For a cross-pollinated plant, the pollen must come from another plant.

Flowering plants are dependent on a host of pollinators, both physi-cal (wind) and biological (insects, birds, and mammals, primarily). Here again it is possible that changing phenologies might result in mis-matches, favoring some species but leading others to decline and even to extinction. An awareness of pollinators and whether they are present during flowering, and in what numbers, is worthwhile, and such tidbits of information are good inclusions in your notes. Somewhat more diffi-cult to discern but also worth noting, if you suspect them, are changed flowering times for species of plants that share a pollinator.

In vascular plants, fruiting occurs at different times depending on whether the plant grows in a temperate zone or a tropical one. Temper-ate fruiting occurs in late summer and early autumn in temperate re-gions, and the time it takes for the phase to complete is about a month and a half (with a lot of variation on either side of that, depending on the size of the fruit and other factors). In tropical regions, different plants fruit at different times of the year.

Some species of flowering (and fruiting) plants do not produce fruits every year. Sometimes they skip several years, and then all the plants of a given species fruit in the same year. The pattern is often unpredict-able. Botanists call this masting, and they are at something of a loss to explain it. Are there cues that tell the plants to skip a year? Or is it an absence of cues? One problem with learning more about masting is that it's a little like Frederic in the *Pirates of Penzance,* who would be released on his twenty-first birthday but who was born on February 29 and therefore celebrates few birthdays. A full phenological study demands a sense of duty. If you become aware of masting in your dooryard, please considering watching it year in and year out, whether the plants are fruiting or not, for as long as you can, keeping good notes.

It is edifying to grow attentive to community patterns of flowering. Just as, in an orchestral piece, the violins all seem to be doing something (even if the first violins are doing one thing and the seconds another), so, too, will you find that groups of plants flower at the same time, some for longer periods and some for shorter. In some deciduous forests, for instance, the flowering plants in the understory rush to get their reproductive business done before the canopy closes overhead, placing flowering plants (and others, such as ferns) in full shade for the rest of the growing season. A few other plants, taking advantage of available niches, may do something different. Be alert for them, too.

PHENOTYPIC PLASTICITY AND OTHER ODDITIES

How well an organism adapts to changing climate depends in part on the degree of phenotypic plasticity it manifests. Phenotype is related to genotype in this way: genotype is a recipe, let's say for bread, calling for yeast, flour, some sugar, some water, and a bit of salt. In the organism, it's in the genes. In the kitchen, it's in the recipe. Very basic, in this case. The recipe probably calls for some sifting, some mixing, some kneading, a period of time during which the bread rises, an oven temperature, and a baking time. If one follows the recipe as exactly as possible, there will still be variation from one person's loaf to another's, caused by such things as relatively humidity and the accuracy of thermostats. Then there are the realities. A recipe may call for sifting the flour, but many a kitchen these days has no sifter. Flours vary slightly from brand to brand (or so those who buy King Arthur flour insist). Some cooks will lessen the quantity of salt, or sugar, to make the loaf healthier.

In the end, the recipe yields a loaf of bread and not beef bourguignon. It doesn't even yield bagels. As long as the variations aren't too out of bounds, the bread will be edible, nicely pungent when it comes out of

FIGURE 6.4. Deciduous oaks provide Walnut Creek Canyon in Arizona, once populated by Sinagua Puebloan people, with subdued fall color.

the oven, and likely to provide nutrition and even a measure of joy. The recipe is one thing, and the product—a loaf of bread—is another. And so it is from genotype to phenotype. But here is where plasticity comes in. If you change one variable, let's say baking time, there is a range in which what comes out is an edible loaf for bread. A few minutes short of a full baking time will produce a loaf that it doughy in the center. A few slices will be fine, but a few won't be edible. On the other end, baking for two long will produce a dry loaf that isn't very good, and even more time yields a burnt loaf. So there are limits to the phenotypic plasticity of a bread loaf. And bread is about midway in the spectrum of recipe plasticities. At one end are scrambled eggs. The number of eggs hardly matters, you can put too little milk or cream into them or too much, they can be a bit runny or too hard. Still, they are scrambled eggs. Hollandaise sauce is a different thing altogether. Use an aluminum saucepan (even one that looks like cast iron!) and the sauce will turn green. Adding too much butter too quickly will cause the sauce to curdle.

And so it is with plant and animal life. A degree of phenotypic plasticity means that a plant will adapt to environmental conditions whether there is too much rain, too little, or just enough. Phenotypic plasticity varies from organism to organism. It also varies from one condition, or demand for plasticity, to another. This is one reason, and it is a crucial reason, that it is difficult to make predictions about how anthropogenic climate change will affect individual organisms, much less ecosystems, for there is likely to be much variability in the plasticity of genotypes from one organism to the next.

A surprising group of plants, to anyone from more northern states, is the broadleaf evergreen, a category that includes the California live oak, which has green leaves year-round, but replaces them with new leaves in late winter through early spring. Therefore, even though its pattern doesn't correspond with deciduous broadleaf trees, it has its own phenological pattern, and reporting networks call for dates of breaking leaf buds and the appearance of new leaves.

Recovery from wildfire is an entirely separate category of study for anyone whose dooryard experiences a ground or crown fire. In an ecological sense, in many places that experience wildfires, recovery may be thought of as truly ecological in nature, but with a phenological twist, because plants will recover with adjustments to small and large changes to the timing of phenophases. And since increased numbers of wildfires are an expected consequence of climate change, those ecologies can be properly interpreted as responding to changing climate. Moreover, the intensities of present-day wildfires often have an anthropogenic fingerprint.

Many of us have grown up with an image of Smokey Bear encouraging the prevention of forest fires, which is a reasonable policy to follow when camping or enjoying picnics in woodland places. But the policy of fire suppression in national forests and national parks has proved controversial. Before the twentieth century, many landscapes saw more frequent fires with lower temperatures and impacts; in many cases, fires are necessary for plant reproduction and recruiting. So if you can include a place on your phenological trail where a wildfire has just occurred, take careful notes and photographs of new life and new growth as it shows up. If the fire was on state or federal land, it's possible that there is already a network in which you can report your observations, if citizen science suits you. Where wildfires were exceedingly hot, a survey of perennial plants, such as live oaks, will show that some survived more or less intact, while others seem to have succumbed to the fire. Many of the latter may have been "topkilled," meaning that their root systems remain intact and they will put up new growth in time— new growth that you may witness, record, and report. Overall, recovery from wildfire is nature going about its business, but to the eyes of a phenologist it can appear, over time, to be a miracle.

Something to watch for is herbivory, the consumption of leaves by herbivores, from insects (at various life stages) to large mammals. The greatest period of damage to plant leaves from herbivores tends to happen early in the season, when leaves are young. Even if reporting forms for citizen science networks do not ask for it yet, it's worthwhile to notice the percentage of leaves, both per leave and the total number, that have been consumed by herbivores. Here, it is mismatching over periods of years that matters, rather than an overdrawn obsession with herbivory. Plants are at the low end of the food chain. Herbivory is natural.

GYMNOSPERMS

A gymnosperm is a seed-bearing plant that does not flower. Best known are the conifers, but others include gingko. Conifers grow across the United States and are the dominant trees and shrubs in many places,

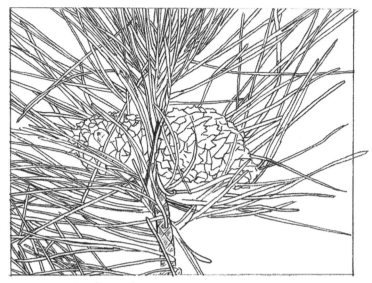

FIGURE 6.5. Ponderosa pine cones.

such as mountain forests, especially in the West. There, they are endangered by drought (generally) and anthropogenic climate change
(more specifically). Especially in the case of wildfire, forests may not
recover and will be replaced by trees and plants from the adjacent life
zone downslope. For instance, ponderosa pine forest will be replaced
by juniper and pinion pine. Elsewhere, stands of pine may be replaced
by chaparral.

Pines, firs, and other evergreens reduce their photosynthetic activity
in winter months but do not cease completely. In order, the events that
one should note and report are: first needles (which are a brighter green
than mature needles), first pollen, full pollen, first cones, and full cones.

OTHER PLANTS AND NONPLANTS

Fungi. Mushrooms are the Pluto of the plant world. Not a planet. Not a
plant. Both plant and planet once, and now no more. Mushrooms *look*
like plants in some ways, and pop up in the vegetable aisles of grocery
stores and greengrocers, but in spite of appearances to the contrary,

taxonomists currently classify mushrooms with other fungi as a distinct biological grouping, seemingly with all the rights and privileges of the plant and animal kingdoms. From a phenological point of view, however, mushrooms resemble plants enough to be included here and not cast out into a (very short) chapter of their own. Why aren't they considered plants? It is because mushrooms, like other fungi, are genetically more like animals than like plants. Because of this, their cell growth and function are so unlike plants that classifying mushrooms with the plants was a kind of genetic nonsense.

For phenological purposes, though, mushrooms are sufficiently plantlike to be considered here. Mycologists, the biologists who study mushrooms and other fungi, use the language of plants to describe the life cycles of mushrooms: they fruit. That's the period when mushrooms look like mushrooms. (The rest of the year their anatomy is quite unmushroomlike.) It's the time of fruiting that may have phenological significance for monitoring climatic changes in landscapes.

Phenological studies of fruiting times for mushrooms (roughly, the time when they appear mushroomlike in the field) go back to the 1960s in Japan, Norway, and the United Kingdom. A number of mushroom species fruit at different times, measured in days, than they once did, but it is not clear whether this is a direct response to temperatures, or whether it is a consequence of earlier growth stages in host plants.

One great value of mushrooms is the way that they tend to stand out from the background at just the time you would want to make a phenological note, as though (not being plants) they aren't subject to any observer's plant blindness. They are treasured by cooks but difficult at first to identify (some mushrooms are poisonous, even deadly), so mushrooms are a keen and challenging object of study for many. Attending to mushrooms when they appear on your phenological trail is worthwhile and quite exciting; after a year or two, you may learn both to distinguish common species and to anticipate the appearance of the aboveground parts, which makes for especially accurate phenological records.

In Europe, phenological studies of mushrooms show that fruiting is occurring later in the season as a response to climatic change. But why? The delay may due to clues from host plants, rather than a more direct cue such as temperature.

Here in the United States, morels (*Morchella* spp.) occur across the country and have a distinct appearance, resembling the finials on Chippendale cabinets from the 1780s. They are treasured both by cooks and by bears. Because there are similar and toxic "false" morels, it's a good idea to learn to distinguish between morels and false morels from expert mushroom hunters. Morels fruit early in the green-up and may or may not be fruiting earlier as climates change.

Let the phenology of mushrooms dictate your interest in learning about them. When they fruit, drop everything else and give them all your attention for an hour or two. And then watch for them around the same time next year.

Mosses. During a cool weekend early in March not too long ago, I attended an annual meeting of historians of science in Friday Harbor, on one of the San Juan Islands. As luck would have it, I had an apartment all to myself right on the water, where I watched the ferries come and go from Anacortes, Washington, several times a day. What caught my eye, other than the lights of the ferries at night, was the moss, or perhaps they were mosses, which seemed to be the dominant species in the vicinity.

Mosses are true plants, one of the Bryophytes, a division that also includes liverworts and hornworts. But they aren't *vascular* plants; they don't have internal plumbing—roots and stems—to carry water between organs. In the Eastern United States, *Thelia* is the most common moss, growing on tree trunks and limbs (living and decaying) and on rocks. Mosses are far more diverse on the West Coast, especially in the Pacific Northwest. Spanish moss, found in Southern states, is not truly a moss (but I discuss it on page xx). If there is moss in your dooryard, make notes on its appearance. This would be a good time to get out a magnifying glass for a close look. Distinguish between different genera or species of moss, and then watch for reproductive growth, which might involve the appearance of stalklike growths and "capsules" at their ends. Make a note when they appear, and when they seem spent.

Sphagnum moss (*Sphagnum* spp.) is the best known of the true mosses; it is usually found in low-lying terrains where water collects but may be dry for parts of the year. Climatic changes bring greater or lesser volumes of water to sphagnum habitat and may create challenges for this moss.

Ferns. When you think of a forest, your mind may not necessarily go to ferns. But think of ferns, and you'll probably think of forests. That's because in the wild, ferns are likely to serve as an understory for a forest, selectively choosing with the shadows of their leaves which seedlings become saplings and, therefore, which seedlings become trees. But ferns can be found in a variety of ecological zones, apart from forests—even in arid environments. Their adaptability has assured that they succeeded through three eons in the geological timescale, arriving with fish but before other animals with backbones.

Ferns are vascular plants that reproduce through spores and do not have flowers or seeds. Because they are vascular, they look more plant-like than the bryophytes; they have roots, stems, and leaves. But without certain obvious characteristics, like flowers, they may be difficult to track phenologically. That didn't stop Alfred Hosmer, a next-generation Concord resident who emulated Thoreau and kept records of fern phenology, based on spore production in ferns. Elizabeth Ellwood and Richard Primack, of Boston University, working together with Jeffrey Dukes at Purdue, were able to use Hosmer's records to show that two ferns living in seemingly similar ecological niches and shared habitats, as well as having similar appearances, at least to people who don't know their ferns (among whom I currently count myself) have varied responses to climate change. Cinnamon ferns seem to adapt more readily to new climate conditions, while royal ferns tend to be flummoxed by them.

Following fern phenology is challenging. You must discover when the spores form, mature, and disperse; this involves a full exploration of the botany of ferns. But currently, the phenology of ferns is not well understood. Climatic change may complicate future research, or it may—with enough data from fern watchers—bring new knowledge to light.

7

Wriggles, Buzzes, and Calls

The tallgrass prairie of North American is not flat, at least not uniformly so. Much of it is dissected, as geologists put it, by tributaries of the Mississippi and Missouri Rivers, making the landscape feel as though it gently rolls. Ever so gently. Neither was it covered at one time entirely by grasses and forbs. There were plenty of trees here as well, although not much more than a fourth of the land was woodland when settlers from the East began to arrive in the early decades of the 1800s. Since then, towns like Carlinville, Illinois, have seen changes in landscape, with large parts of the original prairie tilled for agriculture and others simply mowed on an annual basis. Forests have been cut back and then reclaimed their territories, depending on the ambitions and energies of the landowners into whose holdings they sank their roots.

The grasses and forbs interest us chiefly, though, and particularly the lifeworld without which they could not reproduce. For Carlinville's ecology depends, quite deeply and intensely, on bees, and on the abiding associations between flowers and bees. And while there are many places like Carlinville, or nearly like it, in the United States, it was here, in the county seat of Macoupin, that science found its local proprietor in Charles Robertson, who made a substantial collection of bees, representing scores of species, and where he made many thousands of observations of flowers and their flying visitors late in the nineteenth century and early in the twentieth, leading to the publication of *Flowers and Insects* in 1929. In the meanwhile, Robertson went East, briefly, to take courses in entomology at Harvard, then returned home. He taught for a spell (science and Greek; he wrote some of his notes in Latin, as well) at Blackburn College in Carlinville. Students at the college, many of whom no doubt considered Robertson almost laughably stern behind his fashionable mustache, nevertheless valued his depth of commitment and expertise, and they learned field methods by seeking out the sometimes monomaniacal scholar and scientist, out in the fields near town where he could be found making phenological observations and collecting.

Like many biologists of his time, Robertson indulged in experiments of his own devising, transplanting forbs of various species to plots near his home, the better to make daily observations of them. Over the course of his life he banked reams of data, all of which now provide an invaluable baseline for ongoing phenological studies that can be used to pursue new studies in ecological change and climatic change, too. Recent researchers have not found stark changes; many of the bees that Robertson observed and collected can be found in the fields of Carlinville today.

Will this pattern hold? Only a continued effort to make phenological observations will tell. And so it is across the United States, in the few places where there was an entomologist, or some other biologist, having Robertson's zeal and commitment, and the many places in need of such records.

All animals have phenological significance, in principle at least; most have reproductive cycles and life cycles that are genetically linked to climatic cues of one sort or another. Many of them (marine invertebrates, and vertebrates for that matter) are difficult to observe, though, without special skills and equipment. The insects, especially including bees and their kin, butterflies and moths, and many others, are convenient subjects for phenological observations, records, and reports. They are the stars of this chapter, although other phyla from the animal kingdom, especially frogs and toads from the phylum Chordata, make an appearance.

Many species of fishes are phenologically interesting because of their annual migrations. These, again, make appropriate subjects for the studies of graduate students in biology but aren't generally suitable for dooryard phenological observation except in extraordinary circumstances. And so this chapter will march all too quickly through the invertebrates and what some call the lower vertebrates, dwelling a bit on insects and pausing for an inadequate look at a small selection of amphibians.

WORMS

Annelids, or segmented worms, have life history and, therefore, phenophases. They are difficult to ignore when they turn up after a rain. But in many of the places where they are found, earthworms are introduced species. In the northern United States, the places that were glaciated ten to twenty millennia ago did not have native populations of earthworms until the past couple of hundred years. Ecologists have found that the presence of invasive earthworms tends to decrease overall diversity in a forest unit.

It is worthwhile to observe them when they turn up, or when tilling your garden, and there are dichotomous keys that aid in identifying species of earthworms. But anthropogenic climate change is only one of many anthropogenic change agents that have consequences for the long-term integrity of ecological units. This is a vexed issue in which phenology plays a minor role.

Often, however, "worms" are not true worms but larval stages of insects, such as the "inchworm," which is a larval geometer moth.

SPIDERS

As stated earlier, the life history stages of invertebrates are sometimes difficult to observe without unusual investments in time and equipment. A handful of readers may decide to make that investment because they find the larval stages of bivalves or insects fascinating and the as-yet unsettled phenophases of invertebrates hard not to think about. For the rest of us, a single phenophase may be all of most of these creatures we need to deal with. In the summer of the year I wrote this book, I found a tarantula on the walk from my front door to my car, plus more black widow spiders within a few feet of my bed than I cared to think about. Even so, I made notes of their appearance, the significance of which I cannot guess in a single year and may never know.

E. B. White describes the reproductive cycle of ballooning spiders (minus mating) in *Charlotte's Web*. Spider eggs also protect young spiders through their larval stage, so spiders emerge from eggs as "spiderlings" that must molt as they grow.

Currently, the common wisdom is that the ranges of spiders are likely to move northward, and upslope, in such a way that, for any place in the United States, you are likely to find more spiders and larger spiders, and they are likely to be faster on their eight legs than any spiders they displace.

INSECTS

Entomology, a science often found associated with strong agricultural interests, has been a vital field of science in the United States since the middle of the nineteenth century. With climatic changes, much of the accumulated knowledge of insects—how insects interact with agricultural crops and how to prevent agricultural damage by insects—is quickly becoming obsolete.

The passage from egg to adult in insects involves one of two kinds of metamorphosis ("change in form"). The first is simple metamorphosis, which appears to be nothing more than growth, but it isn't. Instead, insects passing between stages of simple metamorphosis must molt, or shed their exoskeleton, because it doesn't grow with them. While passing through the stages of simple metamorphosis between egg and adult, the organism is called a nymph. Each stage of a nymph's life between molts is an instar. Some insects may grow wings after their final molt. Those without wings may simply appear larger but otherwise similar through each instar.

What is complete about complete metamorphosis is the wholesale change in form from the juvenile organism (the larva) to the adult. In complete metamorphosis, the larva may molt and pass through instars, but before becoming an adult, the larva forms a pupa, in which the transformation from, say, a caterpillar (larva) to an adult (butterfly) takes place.

The common names for most insects usually refer to multiple

species and other taxonomic levels. "Mayfly," for instance, is a common name for thousands of species that have similar appearances and life histories. How particular you wish to become in naming insects depends on your reasons for observing them. The USA-NPN currently accepts reports for more than fifty insects; identifications are required for some of them at the species level, while for others (bumblebees are *Bombus* spp.) at the level of genus is fine. Apart from citizen science, it may still be worth the time to learn to distinguish honeybees (*Apis mellifora*) from species of stingless bees.

The Fishers' Flies. When fishing for trout in the Sierra in my early youth, before the introduction of catch-and-release fishing, we used salmon eggs as bait. It wasn't sporting. Then again, we carried Sierra Club cups and other such vessels that we dipped into Sierra streams without filtering. So, this was quite a while ago. I always hoped one day to do penitence for this early unsportsmanlike fishing and learn to fish with hand-tied flies of my own making. Perhaps that day will come. If it does, I will take my knowledge of phenology with me. Fly-fishing is an attempt, by anglers, to mimic the hatches of aquatic insects using imposters made of feathers and thread tied to hooks. Izaak Walton discussed the ruse in *The Compleat Angler* in 1653.

Mayflies were described by Aristotle and mentioned in Izaak Walton. And in Alberta, Canada, Bob Scammel worked out relationships between flowering of plants and hatches of mayflies and other insects that fly fishers imitate, publishing his system in *The Phenological Fly*.

Mayflies, as a group, actually encompass a few thousand separate species under a single common name. Mayflies spend most of their lives as nymphs, but they have an unusual developmental stage that separates them from other insects. Before the final stage of metamorphosis, mayflies pass through a subimago stage (fly fishers call these duns), in which they are not yet capable of reproduction but no longer nymphs. In this stage, they eat, then pass through one final stage a metamorphosis where they emerge as adults, when mayflies no longer eat. Their life is devoted to reproduction, in a sort of flight dance. After mating, the females deposit eggs in a stream or pond and both adults

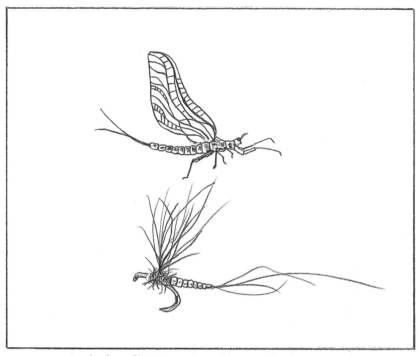

FIGURE 7.1. A pair of mayflies. *On top*, the real insect. *Below* is an angler's hooked proxy for catching trout. Successful fly fishers are often phenologists, whether they know it or not.

die. After emerging from eggs, nymphs pass through one or more dozen instars before reaching the subimago stage. Nymphs that are found and counted in water, whether a stream or a pond, are used as an indication of water quality. Greater numbers point to higher water quality.

During the final two stages, especially the last, you might witness what fly fishers and hikers commonly called a "hatch," even though it is actually a metaphoric change; each mayfly hatched from an egg and passed through the long stages of metamorphosis long before the event you witness. You are likely never to forget your first experience of a mayfly hatch. Mine occurred while I was tracing the path of the Cuyahoga River by foot, canoe, and car from its source (near Chardon, Ohio) to its mouth in Cleveland, years before Cuyahoga Valley National Park protected the river's flanks between Akron and Cleveland. The hatch was north of Kent, Ohio. The air swarmed with mayflies; the road was awash in dead adults (as was the brim of my hat).

The U.S. Fish and Wildlife Service currently collaborates with the USA-NPN to gather citizen science observations of mayfly hatches along the Mississippi River on a stretch through Minnesota, Iowa, Wisconsin, and Illinois.

Like the mayfly, caddis flies actually encompass thousands of species of insects, genetically close and with similar behaviors and appearances. In Minnesota alone, one dissertation writer counted 284 species (from more than three hundred thousand specimens). Unlike mayflies, though, caddis flies go through a complete metamorphosis. Distantly related to moths, these insects create underwater pupae spun from silken filaments. Like mayflies, however, many species of caddis fly emerge at one time, in a hatch, as a "strategy" (although the caddis flies don't sit around a table discussing the best way to confound the rest of nature). Also, like the nymphs of the mayfly, caddis fly nymphs in fresh water indicate good water quality. At any given time, phenology networks, sometimes organized at the state level, welcome observations of caddis flies.

The stone fly is yet another species with an aquatic life cycle. And once again, "stone fly" actually refers to a taxonomic order, numbering thousands of species. As with the other fishers' flies, stone fly nymphs indicate high water quality.

Developing an interest in the phenologies of mayflies, caddis flies, and stone flies can be joined to the pleasures of a sport with history, whether you undertake both interests on your own or join forces with a companion in a form of sport and science mutualism.

Cicadas. If you live in a place where cicadas mate, you know it, may have mentioned it to friends and relatives, and might even have reported observations on social media. The sounds of the males calling to females can be deafening and persistent over weeks. The insects also have a social media following of their own; one website is "dedicated to cicadas, the most amazing insects in the world." Although not universally beloved, cicadas do arouse the sympathies and even love of some who admire them for, among other reasons, their fondness for prime numbers, since they emerge as adults and reproduce over periods of one, two, five, thirteen, and seventeen years.

The common term, "cicada," actually refers to many hundreds of species of these insects. Typically, the females lay eggs in a portion of tree bark, usually on small branches, which die. Entomologists call this flagging, and it is an observable phenophase, worth reporting when seen and confirmed. Eggs hatch in a bit more than a month to three months. The larvae then fall to the ground and burrow into the soil where they feed on roots and pass through molts before emerging as adults, in late spring through summer. Cicadas may emerge in batches. Predation by birds and other animals during the first hatch may be severe, but the din through a period of weeks is testimony to the fact that breeding adults have survived and will produce a new generation.

Beetles are abundant in most of the world's ecosystems, numbering over a third of the million known species. Across the world, beetles are considered pests, although some species, such as the ladybug, are beneficial. Indeed, many beetles are pollinators. Others are primarily herbivores, though, and some consume important agricultural crops. Beetles develop through a complete metamorphosis. It is worthwhile to make observations of developmental stages of beetles when you encounter them.

One well-known example of a set of feedbacks is the mountain pine beetle and its relation to drought. In the West, drought conditions have led to damage to pines from beetles. When there is more water, pines will respond to bark beetle damage by enclosing the beetles in sap. But in conditions of drought, there is not enough sap—the pines are less well defended—leading to deaths among the pines. The resulting deforestation reduces the totality of forested land, and its value as a carbon sink, thereby increasing warming. Whether warming itself is the cause of drought conditions is an open question, but it is more likely to lead to drought than not.

Bees and Stinging Insects. This group of insects includes some that are crucial for horticulturalists and gardeners because of their role as pollinators, as well as some that are considered pests, such as paper wasps. Various groups, often connected to agricultural programs in colleges and universities, have begun or are continuing phenological networks to monitor honeybees, and NASA also has such a program. Even so, notes on honeybees, bumble bees, carpenter bees, and even the first appearances of wasps and hornets should go into your phenological journal, if you keep one, along with dates and notes about weather conditions.

One wasp that I have been amazed to see is the so-called tarantula hawk (*Pepsis* spp.), a frightening critter with red wings that paralyzes tarantulas (but does not kill them) and lays its eggs in the spider's abdomen. Although common in the deserts of Arizona and California, I have seen them in coastal suburbs north of San Diego. The sting from a tarantula hawk is not likely to be fatal, but is said to be extraordinarily painful.

Bees in your dooryard, because of their importance as pollinators and as a source of food for a variety of animals, should always get attention. Bees are members of the taxonomic family Apidae, which includes honeybees, bumblebees, carpenter bees, and stingless bees. Reports of observations to phenological networks are best when you are able to distinguish males from females of the various species. In honeybees, this takes practice distinguishing the slightly larger males from females. In carpenter bees, males and females have different coloration.

Honeybees (*Apis mellifera*), because of their economic importance (both as honey producers and as plant pollinators), are carefully monitored by beekeepers. Even so, it's hard to avoid mentioning them in phenological journals.

Bumblebees are a collection of species, all with large bodies that appear to be fuzzy in yellow and black. The insect has an interesting life history and set of phenophases, beginning with a queen who hibernates through the winter and in spring creates a nest of honey and pollen, sometimes in an existing bird's nest. Here, the queen lays eggs, which she broods until they hatch. The queen feeds the emerged larvae, which enter a pupal stage and emerge as adults. The queen continues to reproduce through the summer, enlarging the colony as she does so. In time, one of her progeny will leave the colony as a queen, then mate and begin the cycle again to replace the colony, all of whose members die, including the original queen. Each of these stages, as well as the identified flowers with which bumblebees associate, are worth noting and reporting.

Unlike bumblebee queens, the queens of eastern carpenter bees (*Xylocopa virginica*) survive to live for two years. Eastern carpenter bees do not eat wood, but make use of cavities in trees and wooden structures to create their nests. Like western, or valley, carpenter bees (*Xylocopa varipuncta*), eastern carpenter bees often seem disoriented, flying in a confusing way that sometimes seems threatening but is not. The eastern bees have an appearance similar to bumblebees. The males of the western bees have a similar appearance, but the females are entirely black. Both bees are pollinators, and reports should include identifications of plants where they are gathering pollen.

Green sweat bees (*Agapostemon texanus*) range throughout the contiguous United States. They are stingless bees. Males have yellow and black abdomens. These are the bees that may gather near picnicking humans, congregating near sources of water, especially sweetened waters such as lemonade and iced tea.

Mosquitoes. The possibility that species' ranges may expand or shift northward has a way of sounding good in the abstract but is less pleasant to contemplate when the species are mosquitoes, especially those that provide a vector for West Nile virus, dengue fever, malaria, and the zika virus. The U.S. Centers for Disease Control (the CDC), which monitors the incidence of these infections, has seen a rise in cases of West Nile virus and dengue fever in recent years. While mosquitoes have been "managed" in parts of the United States for decades, there is rarely agreement about the safety and effectiveness of management techniques and policies. Thus, range expansions and shifts are certain to introduce new management methods, such as trucks equipped with foggers daily patrolling residential neighborhoods, to places that have had little use for or experience with them in the past.

Of course, expansions and shifts in the ranges for mosquitoes do provide new niches for the predators that make use of them for food, such as fishes, frogs, birds, and bats.

As with so many other insects, what we call "mosquitoes" are actually many species. Most develop through a complete metamorphosis, from egg to larva to pupa to adult. While not all females place their eggs in water, most species depend on water as habitat for larva.

The USA-NPN does not presently make requests for reports of mosquitoes.

Moths and butterflies. The novelist Vladimir Nabokov was a curator of butterflies at the Museum of Comparative Zoology at Harvard between 1941 and 1948. While there, he developed a systematic understanding of an order of butterflies—the "blues." To be a successful collector, Nabokov had to know the phenologies of butterflies, even if phenology was not his central concern. In one passage of his autobiography, *Speak, Memory*, Nabokov describes the feelings of ecstasy he experienced on a collecting trip to Colorado, in a "paradise of lupines, columbines, and penstemons."

"I confess I do not believe in time," Nabokov wrote, speaking as a collector and a systematic lepidopterist, in an unconscious (one imagines) counterpoint to Thoreau's "Time is but the stream I go a-fishing in."

But in spite of Nabokov's demurral, he was an accomplished phenologist, attuned to time and the phenophases of butterflies and their surroundings, even coining the term "nymphet" in his best known novel, *Lolita*, from the term for a butterfly life stage.

The life cycles of butterflies begin with eggs, which may be found on the leaves of plants that were selected to provide food for hatched larvae in due course. Inside the egg, the embryo develops to a larval stage, which, in butterflies, is called a caterpillar. Initially, the caterpillars are very small, but they are indefatigable consumers and grow quickly, molting several times as they do so. At the end of the larval stage, the caterpillar forms a pupa or chrysalis. Inside the pupa, the organism goes through metamorphosis, emerging at the end of the process as a butterfly—folded up, at first, but in a few minutes or hours, a complete, and what we think of when we think of an adult, butterfly.

Studies of more than fifty species of butterfly in the Northern Hemisphere show that their adjustment to anthropogenic climate change fits the general prediction. Butterflies are generally migrating and establish ranges north of their former ranges and are moving upslope.

All butterflies are members of the order Lepidoptera. To tell them apart, use a field guide or a dichotomous key. Or study entomology and become a lepidopterist. There are over a hundred thousand species of moths and butterflies worldwide, thousands of which live in or pass through the United States each year. Some are found in some places, or are more easily identified, in their larval stage (as "worms" and caterpillars) while others appear as winged adults. The USA-NPN currently requests reports for more than thirty species of moth and butterfly.

Moths range, in the minds of humans, from pest to pleasure to economic producer. Some moths eat wool sweaters. Others spin silk. Gypsy moths have been responsible for the decline of the boreal forests of the American Northeast and attempts to manage them have led to a decline in luna moths.

In the cases of both moths and butterflies, there are clear phenophases, opportunities for observations, and consequences of climatic changes. The following are a sample of species that can be observed, either as caterpillars or as adults, or both.

The luna moth is large (with a wing span of three to four inches), green, and rather unmistakable in appearance. Its range is to the east of the hundredth meridian, primarily in places where there are substantial stands of deciduous trees and where leaves are not raked and destroyed, a process that also disturbs cocoons. Caterpillars are green, but have brown heads. Just before they enter the pupal stage, they become all brown. Luna moths reproduce once per year, in late spring, in the northern portion of their range; twice to the south of that; and three times per year in southern states. Most observations of luna moths are worth noting and reporting.

The robin of butterflies, the cabbage white (*Pieris rapae*), is a reliable harbinger of spring. They range throughout the contiguous United States and Canada. Perhaps the perfect butterfly to draw with only pencil and white paper, they have dark spots and tips on each wing, although the exact coloring varies. Their wingspans are slightly less than two inches. The caterpillars are often found on the leaves of garden vegetables, such as cabbage and broccoli. Cabbage whites are an introduced species in the United States. Reports should mention plants with which cabbage whites are associated.

Sooty wings (*Pholisora catullus*) are beautiful, darkly colored butterflies with tiny white spots toward the edges of their wings. They are a petite butterfly, not the smallest but diminutive, with a wing span of less than one and a half inches. Sooty wings prefer edge landscapes and meadows. They spend their winters as fat caterpillars, generally enter their pupal stage in midspring, and emerge as adults in late spring. If you observe one or more, make a note of the forbs with which they are associated.

The eastern tent caterpillar moth (*Malacosoma americanum*) has a large caterpillar, often two inches in length and often found in orchards and edge environments in the Eastern states. The pupal stage is easily observed, as eastern tent caterpillars act as a group to create a fairly large cocoon in the forks of tree branches, leaving them to feed. The caterpillars can be quite destructive in orchards.

The western pygmy-blue (*Brephidium exilis*) is a tiny butterfly, with a wingspan of less than an inch and velvety-looking wings that are blue, copper-brown, and yellow at the edges. Found year round in Western deserts, the insects are summer and fall residents in nondesert biomes in Western states. Reports should include identifications of plants with which they associate.

The rosy maple moth (*Dryocampa rubicunda*) displays neon colors: pink or magenta and yellow to pale green. These moths range throughout the Eastern states and are generally associated with species of

FIGURE 7.2. Relative sizes of some moths and butterflies. *Clockwise from upper left*, western pygmy blue, cabbage white, common sooty wing, rosy maple moth, and luna moth, *at center*.

maple. If you see one, the excitement will be enough to encourage you to note it or report.

Ants may seem to be everywhere, always. They aren't. They pass through phenophases, as do other insects. But their often well-ordered activity makes them fascinating to observe, as Edward O. Wilson, of Harvard University, discovered early in his life. Together with Bert

Hölldobler, Wilson spun a systematic work titled *The Ants*, which is worth finding at a library to peruse for a sense of the diversity of these seemingly ubiquitous insects.

Red harvester ants (*Pogonomyrmex barbatus*) are red ants with a range in the Southwestern states. They are sometimes confused with fire ants. Red harvester ants are important in their habitats for their efforts in dispersing the seeds of plants, on which they feed but which they also overharvest. The ants are a food source for horned lizards.

Winnow ants (*Aphaenogaster rudis*) are another seed disperser that range through much of the Eastern United States. These ants, along with red harvester ants, aren't mere consumers of seeds but are engaged with what ecologists call mutualism with the plants on whose seeds they dine, contributing to the reproductive patterns of the plant species. There is some concern among botanists and myrmecologists (ant experts) alike that the phenophases of winnow ants and those of the seed producers may become mismatched. Phenological studies of these interactions are just beginning.

These and other ant species are likely to attract more phenological interest in coming years.

Collecting. Amateur naturalists, including Charles Darwin, have enthusiastically collected insects for at least two centuries and continue to do so, although there are a few restrictions governing collecting on public lands and collection of threatened and endangered species. For many environmentalists, the practice seems barbaric, and the logic required to enjoy (with respect to insects) what you would otherwise abhor (in the case of, say, birds) is difficult for some to follow.

It is likely that a period of collecting will hasten the journey of any citizen entomologist ascending the steep learning curve as they endeavor to learn their insects.

The equipment for collecting insects has changed little over the decades—the basics are a means for capturing insects (a net, a trap, or one's hands), a technique for killing insects without destroying them (usually, a "killing jar" equipped with cotton balls soaked in poison), one or more books for identifying insects, and some pins, labels, and a board for organizing and displaying them. Before beginning a collection, be sure to research current restrictions on public lands. And remember that any specimen must be associated with metadata—date and place of collection—to have value.

FROGS AND TOADS

With frogs and toads, we leave invertebrates behind and pass into the vertebrates. Frogs and toads are amphibians in the order Anura. Phenologically, they are very sensitive to environmental cues such as water and temperature. They are also sensitive to other environmental inputs. Their breeding period is about four to eight weeks in length.

Phenological observations of anurans are made by learning their calls, learning to distinguish between them, and then listening for their calls during breeding seasons, which may number more than one per year, depending on the species. If you've never done it, the idea of learning to distinguish species by their calls may seem daunting, but human ears and memories are more than adequate for the task. It may be somewhat more difficult to separate calls if you hear multiple calls at the same time, a common occurrence. Many frogs and toads are also seen, and not just heard, as adults and as tadpoles. It is possible as well to observe the eggs of many species in ponds and swamps.

Frogs begin their lives as eggs, wrapped together in gelatinous goo. Once fertilized, the embryos develop through a larval stage, sometimes with some parental care, until they emerge either as tadpoles or, after metamorphosis within the egg (in a few species), as fully formed frogs. Frogs that undergo metamorphosis outside the egg have gills and the characteristic tadpole tail at first. In time they develop forelimbs and lungs. Metamorphosis happens quite quickly as they grow to become fully formed frogs.

The spring peeper (*Pseudacris crucifer*) practically calls out "phenology!" in early spring. Well, not really. But the call is distinctive, and generally welcome at the end of winter. A single species with three subspecies, peepers are found throughout the contiguous states east of the Mississippi River and in Texas. As with other frogs, the call is the thing, especially when heard in a chorus. Peepers are found where there are wetlands of any size and begin to call shortly after the last ice of winter has melted. Listen and report the first calls of the year.

American spadefoot toads are a family of toads that prefer dryer landscapes. Their ranges extend throughout the contiguous states, except in the Great Lakes area, parts of the West Coast that see significant annual precipitation, and Louisiana. They are endangered on a state-by-state basis, such as in Pennsylvania. Eastern spadefoots begin calling in middle to late winter. Even though they prefer dry landscape, they call and reproduce at times of significant moisture (rain or runoff). The call of the eastern toads sounds a bit like "Bow. Wow."

American bullfrogs (*Lithobates catesbeianus*) range throughout much of the United States, except for the Rocky Mountains, the Colorado Plateau, the Great Basin, and the northernmost plains states. Within that range, they are found wherever there is still, shallow water. Their calls sound, to my ears, like narrow venetian blinds in a breeze. Bullfrogs call in middle spring.

The name leopard frog actually connotes more than a dozen species of the genera *Lithobates* and *Rana*. The Atlantic coast leopard frog (*Rana kauffeldi*), for instance, is found in wetlands in the New Jersey interior and in Long Island but seems to have expanded its range around Chesapeake Bay and into the Virginia tidewater. All leopard frogs have distinctive spotted coloration (hence the "leopard"). Calls include a fast staccato. The Atlantic coast leopard frog begins to call sometime in middle to late winter.

Boreal chorus frogs (*Pseudacris maculata*) are found chiefly in the northern Great Plains states. They have brown and green alternating stripes. Frogs begin to call in very late winter into early spring. The frogs form a chorus that sounds rather like spring peepers.

8

Feathers and Phenophases

It was a privilege, the first time I saw them. It's a privilege at any time, really. But that's what I thought that first time. A privilege. As indeed it was, even though my sightings were far from unusual. There were six of them altogether, one every thirty seconds or so, gliding a thousand or so below my footing on the south rim of the Grand Canyon. I recalled the way that Roger Tory Peterson described them decades before. They looked to him like bombers from the era of the Second World War—B-17s and B-24s, massive aircraft that lumbered along overhead a life-time ago in places not unlike condor habitat. I looked at my companion, who showed in her eyes that she was in as much awe as I was; although, being younger, she didn't know quite as well as I did what a miracle we were witnessing. Six California condors, riding thermals, going about their business as though their species hadn't been snatched from the precipice of extinction, preserved, bred, and returned to habitat that could support them. In the few short years since that afternoon I've frequently seen condors in the canyon, but I do not yet take them for granted. I hope I never will.

The California condor, along with peregrine falcons and (in the wake of Rachel Carson's warnings in *Silent Spring*) bald eagles, are among a small but celebrated number of success stories, stories of bird species rescued from the fate that befell dodos, auks, passenger pigeons, and (although this last is still not certain) the ivory-billed woodpecker, all of them extinct or presumed by many to be extinct. (Hopeful lovers of bird life simply reserve judgment about the fate of the ivorybill, based on reports of a "sighting"—actually a quite credible report of the call of an ivorybill—in 2004. Many birders consider it bad form to treat the ivorybill as an extinct species.)

Set phenology aside, for just a moment. Seeing condors is worth what little effort it takes to do so, as is seeing and knowing all the birds that make their homes or pass through your dooryard. Of the plant and animal wildlife in the United States, birds have historically attracted much of the spotlight from people who love and care about nature. The National Audubon Society, named for the artist John James Audubon, is currently involved in promoting a spectrum of environmental interests. But it originally had a clear focus on the conservation of birds and bird habitat, as well as the enrichment of its members' lives by the study of birds. It is among the oldest organizations promoting both conservation of and attention to bird life. There is more than good reason to associate the artist's name with birds and conservation. The book *Birds of America* is a masterwork. There have been many times that I've lingered over the elephant edition of Audubon's opus, displayed under glass in the Hawthorne-Longfellow Library at Bowdoin College in Brunswick, Maine. (It is now kept in the library's special collections unit.) But as much as I thrill to the arts of the brush and the printing press, I am all the more thrilled by the sight of a pileated woodpecker in my dooryard, giving me just enough of a look for identification before this red-crested bird circles behind a tree.

Along with millions of others, I was drawn to watching and knowing birds by a Peterson guide, and if Roger Tory Peterson (1908–96) were alive today, I would ask him to write a foreword to this book—for more than a couple of reasons. First among them, I trace my interest in birds to his writings and to other books for which he wrote forewords. More important, Peterson would (I feel quite confident) have understood what this book is about; would have understood the problem the book addresses (climatic change, about which he would doubtless have been alarmed, for his beloved penguins in Antarctica if for no other reason); and would have supported the plea that the book makes (go outside, look at nature, see for yourself if it isn't so, and observe what it amounts to). Peterson transformed the interest of ordinary Americans in nature, Americans like me, to be certain, and perhaps yourself, by making it relatively simple and straightforward to go outside, observe a bird, identify her based on his system of field markings on his pic-

FIGURE 8.1. A pair of pygmy nuthatches. Although not endangered, large patches of their preferred habitat—the ponderosa forest of the American Southwest—may shrink in size in coming decades.

tures of birds, know her name, know her when you see her again. My parents had a copy of *A Field Guide to the Birds*, and although the first copy I bought and owned myself was among the last that he produced (the guide to Eastern birds), and was found to be a mild disappointment by many experienced birders, I still favor it over my much newer Sibley bird guide.

Peterson believed that paying attention to birds, which he and others came to call birding (not bird watching), a special kind of attention that involves not just knowing their names but watching their behavior as well, was a key to preserving nature in the United States. It's something that he said over and over again, a sentiment that hailed from his gut rather than any specific empirical evidence.[1] Before the Peterson guides, a clear bird identification required a specimen: ornithologists shot birds to be certain what they were. Peterson's guides changed that. And while Peterson didn't directly address phenological concerns in his life's work, I have little doubt that he would do so today.

Concern for birds is older in this country than Peterson and his guides. Although the carnivorous among Americans make light of eating a few domesticated birds—chicken is on the menu of almost every fast food outlet—and dine on some water birds with perhaps more of a sense of treading on a wild place, songbirds and several other fami-

---

1 Social scientists have tested the idea and have found that people who spend time observing nature do not necessarily support action on environmental issues as much as Peterson thought.

lies of birds are strictly off the table and have been for more than one hundred years, along with their feathers and eggs. What was once the concern of conservationists with an ornithological bent is now a take-for-granted moral certitude. Would you munch on a bluebird, or fry up a pan of robin's eggs? The loss of several bird species and declining numbers of many species have made birds perhaps the signature conservation interest since the rise of conservation movements in the nineteenth century and of environmentalism in the twentieth. Hats full of bird feathers came to be as unfashionable in some circles in the nineteenth century as furs are today. A concern for birds led Rachel Carson to give her seminal work on pesticides the name *Silent Spring*: silent spring, and no birds sing. For decades, budding conservationists have learned of the plight of the passenger pigeon, the last individual in what was once a vast population in number having died in the Cincinnati zoo. Is there an environmentalist who has not learned to compare any and all threats of environmental degradation to the "canary in the coal mine"?

Passionate birders, meanwhile, have always thought phenologically, whether they knew the word or not. In the crudest sense, most everyone knows that robins herald the arrival of spring in parts of the country. Chickadees spend the winter calling out from under their black caps into the cold air. Canada geese, flying in a V formation, point northward in spring and south in the fall. With greater knowledge of birds, watchers know when to look for the arrival of migratory birds and feel gratified, and a bit relieved, when they show themselves, more or less on schedule. Even when our attentions are drawn to other things, we notice the arrivals of birds, even if we don't make note of precise dates. Early this past spring, when I moved from the Colorado Plateau to a small, temporary residence in the foothills of the Peninsular Range in Southern California, I noticed swallows filling the sky as soon as I stepped out of my car the day I arrived. Who could miss them? Most active near sunset each day, their guano plastered my landlord's stuccoed home in a dappled-over layer of white. And then, on a summer day, I noticed they were gone. All of them. Not a swallow to be seen.

In this latter observation, I didn't capture the exact timing of their departure. They may have been gone for days, or even a week or two, while my mind was on my teaching and other mundane matters. But gone they were, and I noted this in my journal, where the fact doesn't do much to help science but does save for later reflection my sense of the space I occupy. A more precise notation of their departure and—had I moved in time to witness it—their arrival may have had some value. For phenological data about birds tell us two things. First, a phenological record provides an assessment, however small, of the condition of the environment, generally. This is especially true as our planet warms and landscapes change. The second is an assessment of the health of populations of swallows themselves.

And, as Peterson himself so often said (he wasn't alone in this), it's entertaining and edifying to pay attention to birds. Only two days ago, in Maine, I grew attentive as three wild turkeys raced across my path while I drove into town to do errands. They looked rather like dinosaurs. They are, perhaps, dinosaurs. Or dinosaurs, birds. The jury is still out, not so much on the taxonomic relationship between the two groups but about how we should regard this evolutionary fact. Dinosaurs! Crossing my path! But I don't need to see dinosaurs in order to enjoy birds. I clearly recall a visit to Disneyland with family and friends, a couple of decades ago, where I had tired of the rides, the lines, and what for me was a forced front of having fun. I sat at an outdoor table in the center of things, enjoying a refreshment of some kind and marveling at the many hundreds of birds who, no fools they, assembled to take advantage of the environmental edges designed into the landscape of artifice. Birds. At Disneyland.

How are birds attuned to seasonal change and to climate? Where do you look for them, and what do you have to learn? What equipment aids the effort? And how do phenological observations and records of birds enhance your sense of your dooryard, as well as science? Most of this chapter will seem terribly basic to those who have already taken up birding, although some of the tips on specifics of record keeping and paying attention to *changing* nature might be welcome. For readers who

FIGURE 8.2. A mountain bluebird glides over an exotic Western landscape.

have not given themselves over to the joys of birding, the chapter is a foot in the door, one that I hope leads to engagement with the wider worlds of birding, ornithology, and conservation.

The chances are good that your dooryard is a home, a breeding place for migrants, and a crossroads for birds that pass from far away to far away, going one way in autumn and the other way in springtime. There are phenological clues to climate change in all three of these.

Birds can be challenging for phenological observers. It's easy to miss their first appearance, and easier still not to note their last stand come autumn.

Birds are usually busy doing one of three things: making more birds, going somewhere that's good for making more birds, or staying alive long enough to do it all over again. Each of these is conditioned by seasons, especially in the temperate and polar regions of the earth. Over the course of a year, an adult bird may be devoting time and energy to migration (but this applies to migratory birds only, of course), to breeding, and to changing out their wardrobe of feathers (the annual activity known as the molt, or molting). Each of these defines a phenophase in the bird's year. The phenophases do not overlap. Birds don't migrate when they are engaged in reproductive chores (and joys), and they don't molt when their time and energy are taken by the other two activities.

Anyone who loves maps also loves to think about bird migrations, even imagining that birds somehow carry maps tucked under some set of feathers as they make their way north across the United States in springtime or head for wintering places in the fall. Canada geese (*Branta canadensis*), in their precision V pattern, must have a navigator or two who has memorized the map and carries it in her head, landing—as so many migratory birds manage to do—in the same handful of acres each year.

Or maybe it's GPS, in a bird-reader edition. Whatever it is, it amazes, as does the timing. Our Canada goose, the one with the map, seems also to have with her a companion carrying a thermometer. Together, their advance across the country in formation with other geese occurs at a steady thirty-five degrees Fahrenheit. In the 1950s, Frederick Charles Lincoln mapped the migrations of Canada geese and other birds according to isotherms—lines of equal temperature. The black-and-white warbler, he wrote, takes fifty days to cross from the winter range in Mexico and Central America to breeding grounds in the upper Mississippi and Ohio River drainages and the Great Lakes. The gray-cheeked thrush covers a similar distance in about half the time.

They follow itineraries with admirable precision. They anticipate. Birds get ready, growing restless, in many instances, as time for the migration approaches. And they bulk up for what's to come. Throughout the winter they have accumulated fat to provide energy through the migration, especially in bird species that travel across oceans and seas.

Obvious as it seems, it's still worth contemplating the difference between the phenophases that birds respond to and the green-up phase in plants. A plant takes its cues from one complex set of conditions and grows its leaves. For birds, variations in trip timing are the result of a more complex reading of weather patterns. In addition to the bird with the map and the one with the thermometer (I'm being facetious, of course, to say nothing of indulging in the sin of anthropomorphizing; and all the birds have these capabilities), there's one who can read wind patterns and carries a barometer. Many birds like high-pressure air masses for their spring migrations and like to stay toward the west side of such masses or in the center. This makes sense: the clockwise airflow around high-pressure air masses speeds them on their way north, reducing the amount of energy it takes to make it from winter homes to breeding ranges. Birds avoid the opposite sets of conditions, for instance the counterclockwise winds that flow around low-pressure cells, which would impede progress and require greater expenditures of energy. But they do like the west sides of low-pressure systems for their southerly migratory flights as days shorten in late summer through autumn.

Migrations, to put a point on it, are dependent on complexities of weather. Migratory birds move to regions that give them reproductive success and then return to places that allow them to conserve energy through the winter. Changing weather patterns may spell difficulties for migrating birds, possibly tempting some to remain in their breeding range longer than is perhaps wise. My earlier focus on the difference between astronomical seasons and ecological seasons may seem, at first blush, the ravings of a pedant. I assure you that they are not. The dominant phenological cues—the events that plants and animals' respond to out of genetic programming—are day length and temperature. There are other environmental cues, but temperature is the most important. And, as we have seen, temperatures (the environmental cue) are changing whereas the lengths of days (the astronomical cue) are not. This is certain to cause innumerable mismatches in interspecies dependencies.

Here, the value of phenological records of breeding birds is obvious. As climate changes, and with it the seasons, birds will respond in complex ways. Exact dates of migration in spring and fall provide important information about what may be happening.

Migratory birds are among the many sorts of animals (and plants) that cease reproduction in times of colder temperatures, shorter days, and scarce food resources. To put it bluntly, their gonads go south in winter whether the birds do so or not. When they return to their breeding areas, the gonads of both sexes re-form, testes in males and ovaries in females. Temperatures are a factor, but not the only factor, in testes formation. Ovary development responds to a nuanced soup of factors, such as mates, materials for nests, and habitat.

Birds pass through an annual breeding cycle in which they establish territory and mate. The female lays an egg; the egg is kept warm by one or both of the parent birds (or by some other bird or birds). The chicks hatch, are fed, and then fledge, depending on instinct and a learning style informed by grit to achieve flight. Finally, the young birds become independent and the adults breed again or call it quits (their gonads shrink back into irrelevance), depending on species and conditions. Shortening days seem to be a major determiner of the end of the breeding cycle.

The variation in how to accomplish this short list of tasks, from one species to the next, is immense. Territories become established through singing and displays. These latter, along with mating rituals, are a staple of nature videos—the dancing and strutting are potent metaphors for human sexual practices.

Territories may be roughly the same for mating and for nesting, or they may be in different places. Some shorebirds, among others, nest together densely, but rear their young in less dense territories. Birds nest in trees, under the eaves of houses, in stony ground, on beaches. An example of these last is the interior population of least terns, the smallest of the terns, a shorebird that breeds in the region of the rivers of the American interior—the area drained by the upper Mississippi, Missouri, Ohio, and Red Rivers and the Rio Grande. They nest in dense territories on sandbars, on the gravels of glacial moraines and rooftops—any place close to sources of small fish. In the fall, least terns migrate to areas around the Caribbean and the Gulf of Mexico.

Resembling the common tern, but smaller in size, least terns are minimalistic nest makers, crafting nothing more than hollowed-out spaces dug into sand or gravel, which the eggs and the hatched chicks resemble in color. Indeed, this camouflage determines, to some extent, the success or failure of breeding. As are many birds that are reared on the ground rather than in nests above the ground, the chicks are precocial—ready to be about their business a few days after hatching— although parent birds keep an eye on them and feed them as they grow.

It should be no surprise that, given their habitat preference, least terns are an endangered species. The number of activities that humans engage in along the shores of lakes and the banks of rivers is endless, and each one threatens nesting habitat. Changes in climatic conditions are fairly certain to bring changes to lakes, rivers, and water tables in the American interior, although concerning what changes, and with what effect on least terns, nothing is certain.

When their reproductive duties end, birds molt. Without the molt, their feathers would wear out. Changing them—letting go of the old and growing new—makes for a nearly brand new bird over the course of a year, but it requires a lot of energy. So not only do birds tend to do it when they have the energy to do so, they take their time, too— between six and twelve weeks for songbirds. That's up to three months for a wardrobe change. The speed at which the change occurs seems to be regulated by the length of the day. Birds linger at the task when days are long and get with the program as days shorten. Since day length is independent of weather conditions in a warming environment, this provides a mismatch under conditions of climatic change, although its significance is thus far unknown. A variety of birds have yet another change of plumage in advance of breeding, one that is done to promote breeding success. And the extent of the molt, whether it is of all the feathers including those involved in flight (the primary and secondary feathers that provide the wing's trailing aerodynamic shape, plus the tail feathers) or all but those involved in flight depends on the age of the bird. One need only think of the migratory warblers, which have one appearance in spring and quite another in fall as they pass through American's dooryards, confounding birdwatchers' efforts to distinguish between them. In the pre-breeding molt, there may be more concern among ornithologists, perhaps, that phenological mismatches could occur, although warming temperatures during the molt could place birds in greater jeopardy from predatory species.

For more than a century, the greatest threat to birds in the United States has been the loss of habitat, as the nation's human population has increased and modern transportation has permitted, even promoted, settlement at considerable distances from city centers. Habitat loss comes hand in hand with development—loss with progress—but houses and roads are only part of the problem. Development creates new concerns, such as the suppression of wildfires in places where fires have shaped habitats. Look to the Florida scrub jay, for example. The habitat for the scrub jay is now less than 20 percent of what it was in the 1800s, thanks to fire suppression accompanying population growth and sprawl in the state. For this bird and many others, loss of habitat is the greatest threat, much of the time—but climatic change makes matters considerable worse in many cases. This is especially true for the Florida scrub jay, as sea-level rise leads to new losses of habit in the small percentage of their original range that remains.

While the Florida scrub jay's range has contracted, other birds have expanded their population numbers with development and will continue to do so as climate changes. This seems like a positive effect, but often it is quite the reverse—one species' expansion can lead to another's contraction. Many a conservationist developed her early sensitivity to habit loss by watching one species crowd out another.

Birds have long provided models of moral rectitude and depravity alike. In the nineteenth century, introduction of European birds was promoted with enthusiasm, and then decried when the birds appeared to compete too successfully with native birds. Sometimes, the language used by naturalists and nature writers were indistinguishable from nativist writing about human immigrants from Ireland, southern Europe, and China. The English house sparrow was such a species, introduced to North America and then accused of displacing native birds and "fouling their own nests." But there may be no better, and rather older, example of purported depravity than the behavior of the European female cuckoo, which lays her eggs in the nest of another bird, giving us the word "cuckold." The Florida scrub jay (*Aphelocoma coerulescens*) appears to be at the other end of this moral spectrum. Unlike the cuckoo, the Florida scrub jay displays not just a family interest in raising her young but also shares responsibility for breeding with "helpers." Apparently, it takes a village to raise scrub jays. Alas, morality and biology are seldom well matched; scrub jays eat the eggs of other species and are as omnivorous as any human gourmet.

In spite of habits of community breeding aid, habitat loss takes a toll, and populations of the Florida scrub jay were down to about ten thousand individuals in the 1990s, which include four thousand breeding pairs. State and federal listing of the scrub jay with other endangered species brings attention to the species, but what is needed is habitat—and connections between habitats. The habitat of the Florida scrub jay is a case study of fragmentation in the landscape. Once distributed throughout the middle part of the state, the range has been sliced and diced by development, leading to the current arrangement of small patches of habitat, scrubland with oak trees, which must have ground fires about once in twenty years to produce the food sources, such as acorns, that the scrub jays eat.

Far from seeing contraction in their range, northern mockingbirds (*Mimus polyglottos*) have been expanding their ranges, both breeding and year-round, for over one hundred years. As climate warms, they will doubtless continue to expand both of their ranges northward. As omnivores, mockingbirds do well in places that have access to environmental edges, where a greater species richness serves the bird's needs and tastes. Although long a dominant resident of the once rural Southern states (northern mockingbirds are the state birds of five Southern states), northern mockingbirds make themselves at home in urban areas in the north—especially in open green areas such as backyards, parks, and cemeteries.

With their manifest intelligence and their ability to mimic other birdsongs, mockingbirds are often accepted as welcome additions to dooryards where they were once absent. I still thrill to the sight of one here in Maine, where they are summer visitors yet. But the success of the northern mockingbird comes as a cost to other birds, because mockingbirds outcompete other birds for resources of all kinds and may do so quite aggressively. Climatic changes are likely to exacerbate their spread northward without substantially cramping their southern habitat.

A phenological interest in birds can augment a broader interest in birding, or it can stand alone. For many, birding is a social activity. Organizations that promote birding provide opportunities for interested and knowledgeable people, or those hoping to join their ranks, to engage their interest in birds. There are, for instance, Big Day Counts, when groups of birders observe through a twenty-four-hour period, looking to create as large a list as possible of bird species observed through the eyes and ears. The American Birding Association sponsors Big Day Counts in the United States and has overtaken the National Audubon Society as the leading organization devoted to birds, as distinct from broader environmental concerns. The American Birding Association took over publication of *North American Birds*, an authoritative journal of birders' reports, from the Audubon Society, which published the journal as *American Birds*. The best known Big Day Count occurs as the Christmas Bird Count, still organized by the National Audubon Society.

Since there is so much help to be had, it is perhaps a good idea to develop your knowledge of local birds in social circumstances, so that you can easily identify birds that make their way into your dooryard. Birding with knowledgeable people is for many the best way to learn one's birds.

There are the aforementioned guidebooks to birds, and others, of course. These require patience, as one sociologist of science quite expertly discovered. Guides come in two distinct flavors—those with highly idealized, correctly colored images of birds (Peterson and Sibley are the best known) and others with photographs (the Audubon guides). Neither is better than the other, and which to choose is a matter of taste. To find an identification, you need to start by knowing approximately what you are looking for. The identifications are organized in Sibley, for instance, in a recognized taxonomic sequence that begins with birds found in wetlands and hikes on toward the songbirds. You may, in time, begin calling the latter "passerines." The guides provide clear images and notes on field markings, along with range maps (with images and maps appearing on the same page in Sibley, and maps at the back of the book in Peterson).

In many places, there are local guides and checklists. These are sometimes of the quality but, by definition, not the completeness of the guides I just mentioned. They are especially useful, though, in that they simplify the learning and identification process, since birds that are unlikely to be found in a place are excluded.

And identifying birds is indeed a learning process, and a somewhat challenging one at that—but not too challenging. Birding, from one's

FIGURE 8.3. The gear for a morning to be spent watching birds: day pack, binoculars, a container with cup for coffee or tea, and a book for identifying birds.

beginner efforts through more expert stages, probably stands as a perfect example of what the psychologist Mihály Csíkszentmihályi calls "flow," a condition in which skill is closely matched to challenge, providing a state of energy, concentration, and pleasure. And visually identifying birds is only half the interest. Accomplished birders know many species by sound, as well. Think of it: you know, crudely, the difference between a songbird's melody and the coo of a dove, but with some study it's possible to identify many dozens of species by their calls and songs. The Peterson guide series sells a CD of birdsongs, but the treasure trove of bird audio is at the Cornell Lab of Ornithology and can be accessed over the Internet.

There is gear, too, for those who like that sort of thing (a set in which I include myself). Most useful, and a badge of the birders' art, are binoculars. A good pair have objective lenses that are not too large (the larger they are, the less stable your view unless you spring for stabilizing optics) and not too small (the light-gathering capability of larger-than-opera glass binoculars provides a clearer image. A pair of 7 × 35 or 8 × 50 binoculars is a good compromise. For certain kinds of birding, small telescopes are useful, and these are best mounted on tripods. Many birders like to photograph birds. For this, a digital single-lens reflex camera and telephoto or "long" lens are ideal, but discussion of which are well out of the scope of this book.

As you develop as an accomplished birder (assuming you aren't one already and just reading to see what errors I might have made), you may wish to combine your knowledge of birds with other phenological interests, especially concerning sources of food for birds. Note, for instance, hatches of insects and the behaviors of birds when insects hatch. I marveled, one day, at the sight of a lone black phoebe (*Sayornis nigricans*) in my backyard, standing watch on a fencepost before lunging into the air only to snap to what seemed to be a complete halt, with the click of his or her wings. On investigation, I realized that the phoebe was tracking insects and audibly snatching them from midair.

Experienced birders watch for drunken bird behavior in the fall, when overripe berries ferment. Species of birds such as blackbirds and waxwings feast on the berries and are believed to suffer from the effects of alcohol in their systems. Climatic change may increase the number of bird species who "enjoy" this fall bacchanalia.

Long life lists and large counts appeal to some birders, and there is nothing wrong with that. Other birders prefer to focus on behavior, and for them, the breeding season offers great rewards. When watching birds, keep an eye out for these steps in the phenological cycle:

Birds arriving in summer breeding habitat.
Birds establishing territory.
Birds mating.
Birds nesting.
Eggs hatching.
Parent birds feeding hatchlings.
Fledging.

In some cases, a full repeat of the process—mating through fledging. And then, departure to winter habitat.

Nesting, in particular, appeals to many, for several reasons. First, it provides a brief bit of mystery about where the nest is, if you begin by seeing birds carry nesting materials but haven't yet found the site of the nest. Once that's solved, you can watch the birds make the nest. There is nothing remotely parsimonious about the variation in the morphologies of birds' nests. Every species has a different solution to the common problem of sheltering fertile eggs in such away that parent birds can brood and protecting hatchlings while they feed and grow. Indeed, that's what is at issue: every nest has these two functions. Nests need to be cozy enough for the eggs and the chicks and, yet, sturdy enough to support the parenting birds. Given our human zest for constructing domiciles, there is a kind of instinctive mutualism between birds building nests and humans watching them. It goes without saying that you should do nothing to disturb the birds as they go about their business. This is the activity for which more powerful binoculars or telescopes, mounted on a tripod and at a good distance, are made for.

If it suits you, there is nothing to prevent you from knowing every nesting bird or bird pair on your phenological trail, from becoming familiar with their differing nesting styles, or from watching the full breeding process unfold, for several species of birds.

Something to be especially attentive to are numbers of eggs, numbers of hatches, and numbers of young birds at the end of the breeding cycle. The number of eggs is sometimes difficult to determine, although you could climb a tree a bit away from the nest to make a count—in some cases. Northern orioles build pouch-shaped nests that hide the eggs from view, as do cliff swallows. If you are lucky and devote enough time to observing, you may be able to pinpoint the exact time that the mother bird lays her eggs. Barring that, once you see one or the other of the parent birds brooding, the eggs are there.

Predatory behavior, and parent birds' defenses against it, are events to watch for while the birds brood their eggs and raise their chicks. These have significance for monitoring climatic change from year to year. As ranges expand, contract, and migrate, patterns of predation

may change, whether it comes from other bird species or from predators that are not birds. You will want to note each instance of predatory
behavior, the kind of predator, and whether the predator is successful.

The greatest concern for ornithologists and birders alike are the innumerable opportunities for mismatches between breeding birds and
their food supply, as when birds produce larger clutches in warmer seasons, but some of the insects (and other invertebrates) they depend on
for food are actually decreasing in a warming climate.

More than any other form of wildlife, birds offer a glimpse at the
spectacle of climatic change and its connectedness to the web of life.
A phenological eye heightens an observer's grasp—heightens *your*
grasp—of the many connections in nature and the ways that these are
altered as climates warm and change. Because of our species' interest
in and love for birds, there will be success stories like that of the California condor, successes in rescuing birds from near the point of extinction. But there will be losses, too. Every concerned birder, as Peterson
fervently believed, increases the chances of success and decreases the
risk of bird species vanishing from the face of the earth.

9

Warm Blood and Live Birth

Walden Pond was not wilderness in the 1830s, and Henry David Thoreau knew that. There were people to watch, and he watched with a keen eye.

> Early in the morning, while all things are crisp with frost, men come with fishing-reels and slender lunch, and let down their fine lines through the snowy field to take pickerel and perch; wild men, who instinctively follow other fashions and trust other authorities than their townsmen, and by their goings and comings stitch towns together in parts where else they would be ripped. . . . Here is one fishing for pickerel with grown perch for bait. You look into his pail with wonder as into a summer pond, as if he kept summer locked up at home, or knew where she had retreated. How, pray, did he get these in midwinter? Oh, he got worms out of rotten logs since the ground froze, and so he caught them. His life itself passes deeper in nature than the studies of the naturalist penetrate; himself a subject for the naturalist.[1]

It may seem to be beyond the scope of this book, what philosophers would call a "category mistake," to include phenological studies of humans. But why? As one among many mammalian species, humans are often the most easily observed. And while many of their behaviors are attributable to culture and habit rather than to instinct, these too have phenological interest and may be as useful for watching climatic change as those in many plants and animals.

Humans (*Homo sapiens sapiens*) are found in all forty-eight contigu-

---

1 *Walden Pond*, chap. 16, "The Pond in Winter."

ous states, Hawaii, and Alaska. They display a wide variety of colorations in their hair, skin, and eyes. Bipedal, like most birds but unlike the typical mammal, humans often move through the landscape on foot or using technologies they have devised, such as the "auto-mobile," which is linked in most cases to increased atmospheric carbon. Before the last century, animals, such as horses and oxen, were frequently used as beasts carrying burdens of benefit to people. Humans eat almost anything and have domesticated a wide number of plants and animals to provide a constant supply of food on a year-round basis, using technologies (again) to prevent spoilage during periods of transportation and storage.

Females enter estrus thirteen times a year, or about every twenty-eight days. Males are generally unaware of estrus, and, as females may be as well, sexual behaviors are not well coordinated with reproductive realities. If copulation leads to fertilization, gestation takes about nine months. At birth, an infant human is entirely dependent on the care of a parent (or surrogate) and generally does not leave the parent's or surrogate's care for eighteen years or more. Brain development is slow but steady over this time. Both females and males are biologically ready to reproduce at about twelve to fourteen years of age, but cultural standards generally favor a delay in reproduction for another ten to twenty years.

Human phenophases are not well understood. In the United States, annual curves displaying mortality show a spike at and near New Year's Day, which occurs about two weeks after the winter solstice. In a small percentage of the population, there is a likely causal connection between short day lengths in winter and depression, combined with lethargy. Formal learning is generally a three-season activity in childhood; summers are given over to alternative behaviors of various kinds, not all of them designed to increase rationality. Greater exposure to sunlight may cause a darkening of some skin pigmentations during periods of longer day length, leading to "tanning" in some members of the population and "freckling" in a small percentage. Using technologies that some consider to be the earliest cause of increasing accumulations of atmospheric carbon dioxide, humans "clothe" themselves in materials of varied insulating value and reflectivity during the year.

Humans have varying degrees of awareness of their relationship with the rest of the natural world. In his signature work on the importance of nature for humankind, *Biophilia*, the natural historian and ant expert Edward O. Wilson wrote that "the naturalist is a civilized hunter." An ironic choice of word: "civilized," for Wilson describes his hunter afield, in a place utterly uncivilized, either by other people (the hunter ventures out alone) or by the works of humankind. So, whatever did Wilson intend? Did he mean that his hunter was *peaceful*, as opposed to predatory or warlike? I suspect that Wilson meant by "civilized" a synonym for scientific inquiry as Wilson sometimes practices it (that is, when he wasn't extinguishing all life on an offshore island). I think I understand him to mean that the naturalist is a *zen* hunter, a *mindful* hunter. Still, I'm left wondering. Aren't all skilled hunters at least this, or aim to be?

Wilson's civilized hunter paid attention to the swarming of midges and to the smells in the soil, in order to collect salamanders, unlike the sort of twenty-first-century American hunter of game, who directs attention primarily to mammals and a few birds—and, more than any other mammal, at deer, particularly white-tailed deer.

Phenologists share a variety of strategies with hunters, such as tracking spoor. Spoor include tracks, the impressions that animals may leave as they move through a landscape. There is also scat, or excrement, which has a notable "signature" that varies from species to species. And there is other evidence, such as tree bark that has been marked by a bear scratching his or her back. An awareness of spoor is a more than suitable way to watch the movements of larger, predatory animals such as big cats, bears, and wolves.

There was a time when "civilized" hunters even sought after birds and large mammals with guns to collect them, but that sort of activity is considerably reduced from what it was a century ago. There are other ways to "collect" mammals and birds, and counting is among these.

Joseph Grinnell, first director of the Museum of Vertebrate Zoology

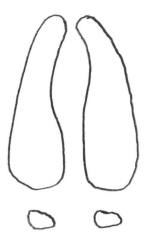

FIGURE 9.1. Although some mammals present themselves for observation more than others, often it is spoor—such as this deer track—that provides the best way to "see" mammals.

at the University of California, Berkeley, was an old-school naturalist, collector, and hunter. Over the passage of four decades, he and his associates pieced together a collection of thousands of specimens to document California wildlife still extent in their own time, as well as wildlife no longer present. The collection includes a California grizzly bear, now extinct.

No mere hunter, civilized or otherwise, Grinnell took care to collect metadata with his specimens and set standards for record keeping as discussed in chapter 4. Today, the collection is a vast portrait, in cabinets, drawers, and notebooks, of Californian life zones as they were in the past, up to one hundred years ago. Much more recently, ecologists and conservation biologists have collected data from many of the same sites where Grinnell, his colleagues, and his charges collected specimens. The result is the Grinnell Resurvey Project, a record of California's changing ecological conditions.

Grinnell devoted much of his career to birds, but his careful records of mammals, especially smaller mammals such as squirrels, mice, and shrews, have provided resurvey investigators with considerable evidence of climate change made manifest in alterations in the sizes, locations, and altitudes of ranges for California mammals.

Phenological observations augment these ecological field studies. More to the point, not every part of the United States had the luck to sponsor the work of a Grinnell. In many places, ecologists have much less to go on, and phenological observations, along with breeding surveys and other data, are essential for tracking the effects of climatic change.

Are mammals helpful in this? Mammals are attention-getters, without a doubt. That said, mammals are in some ways more difficult to observe than other fauna or flora. They are often more secretive and sometimes more dangerous; an encounter with a bear or a mountain lion is doubtless bracing but also threatening and ill advised, if for no other reason than that hunters may be afoot in the ranges of some mammals—armed hunters, some of whom cannot fully distinguish between two-legged mammals and four-legged varieties. The best way to experience behaviors associated with phenophases in large predators is to watch videos. For this and other reasons, observations of mammalian phenophases by citizen scientists and amateur naturalists tend, in most cases, to be dependent on opportunities, not on disciplined observing over seasons or years.

Even so, there are opportunities. Development in the United States has been such over the past seven decades that humans now increasingly dwell in places where there are large mammals, including bears, moose, bobcats, mountain lions, and elk. Deer have been garden-variety mammals for some time. Suburban dwellers regularly report coyotes, skunks, raccoons, possums, and (in some places) armadillos. Common suburban mammals include squirrels, chipmunks, voles, moles, rats, and mice. Wild dogs and feral cats are not uncommon.

Mammals are warm-blooded. Our bodies are covered with hair or fur. The female of mammalian species in the United States give birth to live offspring following a period of gestation specific to each species and suited by natural selection to be seasonally optimal. Fertilization and mating are similarly timed to produce the largest number of healthy offspring. But here, there are some differences among mammals. In some species, a newly fertilized egg forms an embryo that implants in the womb and develops through gestation to birth. In other cases, embryos are dormant for a period of time, implant later, and then develop to birth.

A number of mammals, such as bats and many rodents, hibernate as a way of adjusting their metabolic requirements to diminished supplies of food in winter. Others, such as bears, have similar instincts for surviving winters, although biologists may not class these strategies as true hibernation.

Hibernation and similar strategies are phenophases; they are regulated by astronomical and environmental cues, especially temperatures. Because hibernating mammals are usually not observable without invasive techniques—and are therefore assumed in the absence of observations—some phenological networks do not list hibernation and related inactivity as reportable phenophases.

In true hibernation, animals that control their body temperatures (ectotherms) lower them as a way of conserving energy. Heart rate declines, and respiration diminishes. Torpor is a similar strategy. While hibernation is not easily observed, certain clues that animals are preparing to hibernate are observable—the caching of nuts, for instance, or the development of winter coats.

Because of their place in the food chain, a wide variety of plants and animals (and mushrooms) have phenologies that have consequences for mammals. It is not safe for citizen scientists to engage with grizzly bears, for instance, except to make a note of the occasional sighting—ideally at a distance. But bears rely on huckleberries as a portion of their diet, so scientists with an interest in bears look to citizen scientists collecting data on huckleberries and other food sources. This is a useful way to make a contribution to science. Much the same sorts of observations are valuable even for mammals that are not dangerous. Food sources, whether for squirrels or bears, are likely to change and shift in coming decades. If you glimpse a rabbit nibbling at plants, or squirrels catching insects, try to identify not just the predator but also the prey and watch for change or stability of these relationships over time.

Squirrels, mice, and rats like our company, although they tend to be wary when we pay any attention to them. But since they get along so well with us, we can expect that changes in our life chances will reflect in their life chances as well.

Squirrels are squirrels, but so are woodchucks and prairie dogs.

Chipmunks are squirrels and so are flying squirrels. Tree squirrels and ground squirrels are squirrels. Indeed, there are more than 250 species of squirrels, although only a fraction of them have a range in the United States. The reproductive cycles among the squirrels have many similarities from one genus and species to the other. After my initial discussion below, I will mention only differences.

Let's start small. Eastern chipmunks range from the Mississippi River valley to the East Coast but are absent in parts of states bordering the Gulf of Mexico. Western chipmunks, a separate genus and divided into twenty or more species, are found in the rest of the United States, except in the Sonoran and Mojave Deserts, the Sierra Nevada mountain range, and the Central Valley in California. Eastern chipmunks hibernate in winter, while western chipmunks burrow and consume a winter store of foods. Both species mate in springtime and may mate a second time in fall.

Chipmunks, eastern or western, are worth noting. Make a note of food sources if you observe one or more chipmunks eating or gathering nuts in their cheek pouches.

Tree squirrels alone include dozens of species, best known of which in the United States are gray squirrels and red squirrels. But on the southern Colorado Plateau, the tufted-ear Kaibab squirrels and Abert's squirrels pop up in the ponderosa pine forest on either side of the Grand Canyon, Kaibabs to the northwest of the Colorado River and Abert's to the southeast. The two are considered subspecies, but they are isolated geographically and were once thought to be separate species. Kaibabs and Abert's are dependent on the pinecones of ponderosa pines as well as fungi that live in the roots of the pines, and their survival is linked to the forest, which will have to recruit upslope to survive.

Gray squirrels range through the Eastern United States and the northern Mississippi River basin. Their expanding range extends into southern Canada. They excel at sharing human habitats, especially urban green spaces, where they are often common and observable wild mammals and taken for granted for that reason. Usually gray in color, there is considerable variation in parts of the range. Dark (melanistic) individuals are common in northern parts of the range. Their preferred foods are nuts, seeds, tree bark, and fungi. A more generalized diet, including insects, may signal periods of stress or may simply be a response to broad opportunities.

Mature gray squirrels may breed twice in a year, but when they are younger, the female might reproduce but once in a year.

American red squirrels are woodland mammals with a preference for pine forests. They are found through the Northern states and well into Canada, as well as in the Rocky Mountains. Red squirrels have an unusual mating behavior, in which several males chase a single female. Pine seeds make up a major part of their diet.

Mice of several species are common dwellers in dooryards and wherever humans reside. They are generally considered vermin, especially where crops (such as corn) are stored. Where they are unwanted, such as inside human habitations, especially kitchens, it is somewhat common to keep a predator such as a cat. Mice are favored food for a large number of predators across the chordates: owls, eagles, snakes, lynx, and others. Deer mice have been linked to both hantavirus in the Southwest and Lyme disease, which they get from ticks.

Rats seem to be especially well suited to urban environments. Any major change in the distribution or phenologies of rats would be interesting news, to say the least.

Rabbits and hares. Has this happened to you? You have an argument with a colleague, or the boss calls you on the carpet for some minor error. Or a home repair turns out to cost hundreds more dollars than you'd planned to spend. Maybe you just took an exam and worry that you've not done as well as you'd hoped. Sullen, you head for the nearest greenspace where you can escape the overcivilized world. And as you turn a corner in the path, you encounter: a baby bunny. This happened to me many times on the campus where I taught for many years, and each time it did, I felt renewed. It is very difficult to remain angry in the company of a baby bunny. And even though this setting was also coyote country, the presence of any youthful rabbit is proof that at least two adults rabbits made it safely to the age of reproduction.

Female rabbits (does) are capable of reproducing several times a year, beginning in late winter, mating with a male (buck) over a matter of a few minutes, at most. The gestation period is about a month. The young are born blind and without fur but develop in a little over two weeks and are weaned after about a month. Rabbits are entirely herbivorous, and the cellulose from the foods they eat is digested primarily in their lower digestive tract.

Eastern cottontails are widely distributed east of the Mississippi River, the Great Plains, and the desert Southwest. Desert Cottontails inhabit the western Great Plains and the desert and mountain West. It is worth making note of sightings of cottontails, counting them when there is more than one and paying attention to what they are eating. Although notorious for being prolific breeders, rabbits could experience consequences from climate change.

Hares, while superficially similar in appearance to rabbits, are quite different from them. Male hares (not bucks but jacks) must chase females (jills, not does), catch them, and then fight with them before mating. In hares, the gestation period is nearly two weeks longer than for rabbits, and the young (leverets) are born with fur and able to see.

Snowshoe hares (*Lepus americanus*) are quite interesting, phenologically, as they molt and, in doing so, change their fur color from brown to white, so as provide camouflage in snow cover, then back again when the snow melts. The primary mechanism in these phenophases appears to be day length, although temperature and the presence or absence of snow may play a role—but not enough of a role, perhaps, because there are increasing observations of mismatches in spring and fall in this species. Biologists hypothesize that natural selection will drive changes in populations of hares to accord with shorter periods of snow cover.

Snowshoe hares are found in mountainous areas in the Pacific Northwest and the Rocky Mountains, in northern Minnesota, Wisconsin, and Michigan, in the Appalachian Mountains north of Tennessee, and from New York through New England.

Jackrabbits, white-tailed and black-tailed, range through the Western states. All of the hares have longer life spans than rabbits do and are worth noting when observed. Be sure to record fur color.

Bats. There are around four dozen species of bats in the United States. Worldwide, bats are of two main types: megabats, and microbats or echolocating bats. Only the latter are found in the continental United States, including little brown bats and big brown bats, both common in most parts of the country. Echolocating bats emit sounds and instinctively process the return times of echoes in order to "see" objects, particularly prey, in their habitats. Both are insectivores and are therefore dependent on the phenologies of insects. A group of scientists from the University of Wisconsin, Milwaukee, working with Wisconsin state biologists, studies little brown bats and has found little change in the timing of their spring emergence from caves.

Along with most bats in the United States, little brown bats are difficult to observe and identify. They are active at night and move rapidly. But if you are certain of identification, you can look for first appearances in spring, after the bats emerge from hibernation. Although mating occurs in the fall, before bats hibernate, the females store sperm and are fertilized in spring. Gestation is about two months; young are born in summer.

Bobcats and mountain lions. A ranger in southern California once told me that whenever I was hiking in the region, I could be fairly confident that a mountain lion was watching me. In time, I got used to the idea, but I always feel more confident, if that's the word, when I see mule deer on my hikes: a hiker or bicyclist may be interesting to watch, or at least worth keeping an eye on, but a mountain lion would much rather dine on venison.

Mountain lions, also called puma and cougars, occupy a range vastly reduced from what it was in the United States thanks to bounties placed on them in the not-so-distant past. Although lions will hunt and feed on smaller mammals, they prefer deer and thus inhabit similar habitats, individually, except in the case of a female whose young are still with her.

Jaguars are found at warmer latitudes, with infrequent sightings in the border Southwest.

The lynxes are a smaller species of wildcat, two species of which are found in forty-nine U.S. states and in Canada. *Lynx rufus* can be found throughout the contiguous forty-eight states, with the exception of the areas around the prairie states and the Great Lakes. They are commonly called bobcats. *Lynx canadensis* are primarily a more northern species, but overlap with *L. rufus*. The bobcat is considered a game animal in many states. Bobcats have a diet similar to that of coyotes (but with more success capturing birds), and their habitats may overlap. Bobcats mate in late winter. The females give birth to four to six kittens a bit over two months later, in midspring. Kittens will stay close to their mothers for about a year.

Bobcats are active in twilight, near sunrise and sunset, and one sometimes encounters them on early morning or early evening hikes. While hiking in bobcat habitat, one should be alert to bobcat spoor, tracks, and scat, making notes when you find these. If you are lucky enough to see a cat with prey, be sure to document the kind of prey.

Foxes, coyotes, and wolves. The red fox inhabits a large part of the contiguous United States but is absent from parts of Idaho, Washington, Oregon, California, Nevada, and the Southwest; in most of the range where red fox is absent, one finds gray fox. The red fox is an omnivore, hunting and eating other mammals as large as raccoons and juvenile deer, as well as other chordates. Fox supplement their diet with plants and some invertebrates. They prefer edge environments, and this brings them into urban areas where they forage in gardens and the middens of human populations.

Females (vixens) enter estrus in early winter; they copulate with a male for a period of about an hour, after which there is a gestation period of about two months. When the vixen gives birth to four to six kits, she is usually immobilized in a burrow for fourteen to twenty-one days, and the male feeds her while she suckles the kits. Vixen and kits emerge from the burrow after a month's time and begin to hunt with the vixen and the male fox.

Because fox are wary, especially in areas populated by humans, it is exciting to see one and to jot down their numbers and the presence of kits, if present. The vocalizations of fox are varied, particularly when mating, and worth learning and chronicling when heard.

Gray fox look similar to red fox from a distance; it is best to learn to distinguish them in the company of an experienced observer. Gray fox are found throughout the contiguous states, except in the northern

Rocky Mountains and the northern Great Plains, and are more inclined toward woodlands and other places with cover, unlike the edge-loving red fox. Vixens enter estrus in midwinter and gestate for six or seven weeks.

Coyotes are doing well and are likely to continue apace as climate changes. Inhabiting all of the contiguous United States and Alaska, they were absent from the southern Appalachian states and Florida but have expanded their range over the past half-century. They are broadly omnivorous animals and are among the set of species that thrive in proximity to human settlements, sometimes feeding on pets, such as cats and small or immature dogs.

Coyotes mate in middle to late winter; pups are born about two months later. Both parents play roles in rearing young into early summer, when the pups are weaned. The pups are mature by the middle of the fall of their first year.

Although it is not unusual to observe coyotes visually, particularly lone individuals, it is more common to hear coyotes. If you do happen to encounter a coyote visually, it is worth watching what is being hunted and marking it down. Also note the condition of coat and apparent weight (lean or well-fed).

Raccoons, skunks, otters, and badgers. Raccoons and skunks are among the Musteloidia, a superfamily that also includes weasels, otters, badgers, martens, and close relatives to raccoons such as ringtails and coatis. Of these, raccoons and skunks are most likely to frequent the American dooryard. They are generalists, and the skunk in particular has evolved a formidable system of defense against predators, real and imagined.

Raccoons' ideal habitat includes streams, ponds, and lakes, and it is fair to say that the species was for a long time a free rider in beaver habitat. During the early period of European expansion into North America, many streams were transformed by humans into dam sites, and this as much as anything else contributed to the tendency for raccoons to associate with human populations and may have favored an expansion in the species' range, albeit with a wariness about humans.

Raccoons mate in late winter. The phenophase seems to be controlled by length of day, although temperatures may also play a role. The period of active mating—courting and copulation—occurs in under a week's time. In the aftermath, a fertile female will gestate for about two calendar months, at the end of which she will give birth to a litter of two to five young. At two months the kits begin exploring and will forage; they are fully weaned at about four months. In the fall, kits will go their separate ways, and males roam some distance from the places where they were born.

Raccoons have successfully adapted to a wide range of changes to habitat and will probably continue to do so. Shortened winters are likely to increase the extent of each year that they are active.

Make notes of sightings of raccoons. I was surprised one day to see a raccoon peering out from a storm drain in San Diego. Whether it was merely hiding there until I passed by or was living there, I had no way of knowing.

Skunks, too, are found in a variety of habitats across the United States, including urban areas, where they forage for foodstuffs discarded by human populations. In all habitats they are omnivorous. Adults mate in early spring, and females then gestate for a little over two calendar months. Males do not play a role in rearing offspring. In colder regions, skunks will occupy winter dens but they don't hibernate.

Log sightings, as well as smellings, of skunks.

The rest of the Musteloidea are uncommon enough to deserve reports, in your journal at least, on any and all sightings. Weasels are common in rural landscapes, especially farms and tailored environments like golf courses. They can breed twice a year, in spring and again in summer. All sightings are worth noting. Badgers are omnivores. Their reproductive cycle is unlike other Musteloidea. Badgers mate in late summer or early fall. The fertilized embryo then remains in a dormant state, implanting in early winter and gestating over a period of less than two months. Shortened winters may have an effect on badger populations. All sightings of badgers, and especially of young, are worth documenting.

Deer and elk. Deer are the largest mammal that you are likely to see on a regular basis. I recall the thrill I felt one night, while sitting at a window gazing dreamily at a winter landscape in moonlight, as nine deer of varied ages emerged from the wooded yard across the street, crossed the road one by one, and passed through my dooryard to some-place behind my house.

To my taste, the carefully choreographed chain of deer was a gift. For many, though, deer are the largest nuisance they are likely to en-counter, a menace to successful gardening, a big suburban pest. White-tailed deer range throughout the contiguous states, with the exceptions of the Colorado Plateau, the Great Basin, California, and the Puget Sound. They show a preference for "edge" environments combined with sources of water. For this reason, suburban landscapes bounded by woodlands are quite ideal—and the cause of suburban gardeners' displeasure at seeing deer pose in a well-tended garden space. In the far West, mule deer are likely to occupy similar habitats. White-tailed deer seem to have expanded their range through much of the twentieth century and are popular and abundant game for hunters.

White-tailed deer and mule deer are members of a broader taxo-nomic unit that also includes elk and (in Canada) caribou.

Estrus occurs in female white-tailed deer (does) in the middle of the fall. Length of day seems to determine this phenophase, which is known as the rut. Mating with males is a fairly perfunctory affair; ges-tation is just shy of seven months in length. Fawns, one to three in num-ber, are fed and sheltered by does for about a month, then begin accom-panying the doe as she forages. They wean at about ten weeks. Females are ready to mate in a few months; males usually have delayed repro-ductive success until they are about eighteen months of age.

Because of their place as a managed game animal, white-tailed deer already provide a lot of data for phenologists. Even so, it is worthwhile to note sightings, especially in spring when does may be accompanied by fawns. Years with early springs are associated with better survival of fawns.

In the West, mule deer may migrate from warmer deserts and valleys in summer to mountains and mountain foothills in winter. Reproduction is similar to white-tailed deer, although the rut continues in early to midwinter.

Annual variations in conditions for mule deer in Western states, especially during periods of drought, may cause temporary range expansions of mountain lions, sometimes leading to incidents that bring lions and humans together.

The habitat for elk is primarily in the mountains and plateaus of Western states. Larger than white-tailed and mule deer, elk migrate from higher country in summer to lower altitudes in winter. With anthropogenic climate change, the lowest elevations at which winter elk are found are likely to increase, and elk are likely to migrate to higher elevations in summer as long as there are sufficient sources of food. For elk, the rut begins as early as late summer and continues through the fall. Female elk (cows) gestate eight months or more.

Because of their migratory behavior, notes on sightings of elk are worth keeping. Count numbers; make a note of cows, males (bulls), and young (calves). Very young calves have spots.

Moose are distantly related to deer and elk. Ranging through the northernmost contiguous U.S. states and into Canada, their reproductive cycles are similar to elk. Because moose found in the United States are at the southern extent of their range, it is possible and even likely that moose will be extirpated in the contiguous states in due course owing to anthropogenic climate change. All sightings are noteworthy, and details—numbers, genders, presence or absence of calves, apparent condition—are worth keeping.

Marine mammals. Although smaller than blue whales, gray whales are among the largest marine mammals. Eastern gray whales migrate along the West Coast of the United States, from the Bering Sea to the Gulf of California, responding to phenological cues—sunlight and water temperature. Because their migratory route takes them close to the coast of California, whale watching—usually from private or chartered boats—is a favored life event for naturalists. Other whales are more elusive.

Of the whales' phenological cues, water temperature, in particular, will change as global climate warms and may cause changes in the timing of whale migrations and the availability of resources at migratory endpoints. Without private wealth or a research grant it may be difficult to monitor whale migrations for phenological data, but fortunately the popularity of the migration is such that changes in dates of migrations are likely to be reported by institutions such as the Monterey Bay Aquarium.

Miscellany. Ground squirrels and rodents in the mountains and foothills of California and Colorado have been found to host the fleas that carry bubonic plague. Usually, this occurs on public lands and is likely to be posted when there is concern. Changes in phenology could lead to changes in the vector for this disease.

If your commute is regular it is worthwhile to make notes about roadkills of armadillo, possums, skunks, and the like.

Domesticated animals: cows, goats, pigs. One of the reasons that certain animals were suitable for domestication was that estrus in these animals is subject to manipulation or is repeated within a yearly cycle. In sheep, estrus is seasonal in some breeds, less seasonal in others. Bovine females enter estrus for a period of eighteen hours every eighteen to twenty-four days. In pigs, estrus is manipulated by breeders. For these reasons, domesticated animals are not very interesting subjects for phenological study.

10

The Atmosphere at Home

Just as weather isn't climate, so too it isn't phenology. But the two are so closely related that, to understand one, you do well to better understand the other.

Meteorology and phenology, as scientific pursuits, have certain commonalities. Both rely on data collected from a broad area, which can be modeled in order to provide predictions. Meteorology has been pursued by many more people (its usefulness for commerce, from aviation to knowing when to get out and plow winter roads, is undeniable), with more access to resources, and for a longer period of time than phenology, so that the models for meteorological prediction are far better developed than are those for phenological prediction. In time, this may change. While data collection for weather forecasting and reporting is a long-established form of citizen science, with similarities to burgeoning citizen phenology, the importance of phenological studies and models, developed with access to data from citizen phenologists, is likely to grow.

The weather in your dooryard and the weather as reported by the National Weather Service, or by a private source like Accuweather or the Weather Channel, have a rather close family resemblance but they are not the same. True, they don't vary a great deal from each other—a cold winter day for the National Weather Service is likely to be a cold day in your dooryard—but they aren't clones of each other, either. In a sense, everyone knows this. Complaints about weather forecasts gone awry—forecasts for snowstorms that don't quite materialize, sunny days that bring rain to carefully planned picnics—are legion. But the differences between forecast or reported weather and the weather you experience are more curious than this. The weather in your dooryard

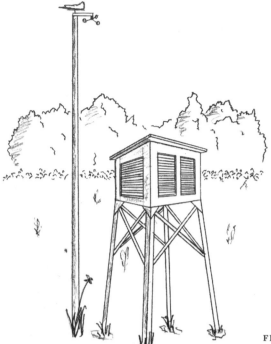

FIGURE 10.1. A Stevenson
shelter for weather instruments.

is many times more complex—and interesting—than the weather you
hear about when you turn on the television to the Weather Channel.

Let's begin by thinking about temperature, which is not just the central data point for anthropogenic climate change but also (as we've
seen) a well-known cue card for phenological events. I've long since
forgotten where I read it, but some wise writer once claimed this: the
organism that experiences the air temperature that gets reported on a
bank sign, along with the current time, is a spider, the spider that lives
in the little shelter shading that particular temperature sensor from direct rays of the sun. The temperature you experience is almost always
different from this, for a variety of reasons.

The National Weather Service, as a rule, shelters its temperature
sensors (thermometers) from direct sunlight. A reported temperature
is a temperature in shade, and it's the temperature of air that is circu-
lating, at least to some degree, in and out of the shelter. It is not the

temperature at ground level but, rather, somewhere above the ground, usually five feet aboveground, plus or minus one foot. In other words, it is a *standard* temperature reading (and the thermometer has been calibrated so as to be standardized with other standard temperatures). Such a fact, this standard temperature, has many uses, particularly for feeding data to those models that help forecasters make predictions of weather in the future.

But instead of predicting it, let's take a contemplative view of weather. Weather is as much a part of our natural world as the rocks, soils, water bodies, plants, and animals it touches. And yet we are tempted by custom to think of weather as a thing apart. We often have good reason to do this. Tomorrow's picnic or hike need not be canceled because of today's drizzle. Monday's commute need not be made more dangerous by Saturday morning snow. Weather transpires and passes away.

That's one way to see it. Snow melts, the ground dries and is wetted again, the sun shines, all in present time. Seeing weather as we see a bloom or a bird on the wing, seeing it as nature, is worth our while. Indeed, looking at weather, rather than through it or past it, is attentiveness at its best. No one knew this better than did the nature photographer Ansel Adams. The crystal clarity of one of his photographs, in which a cutting sharpness extends from the very distant through the very close at hand, complements visible signs of weather in some of his finest work. The signs of weather are themselves clear too, as crisp as the rocks and the trees; despite their clarity, we know these visible signs of weather in our minds to be in motion, fleeting and transitory. They are wisps of time and *in* time. There is no better example of this than *Clearing Winter Storm, Yosemite Valley, 1944*, in which Adams captured the valley as few tourists had then seen it. Clouds choked the eastern end of the valley, obscuring Half Dome, Adams's "monolith," and the bottom half of El Capitan is veiled in a mist of ice.

We can take a contemplative view of weather. We can watch with interest when weather draws its ears back, gets up on its toes, and grows nasty, presenting itself as nor'easters, Santa Anas, hurricanes, and thunderstorms with threats of tornadoes. Under such conditions, what was tame and civilized a day before quickly becomes wilderness—a sort of temporal wilderness, a wild *time* more than a wild *place*. Things that were known and taken for granted suddenly are not.

As I write this, I look over the top of my computer screen, through a window to a pine tree whose branches flex in a strong late-winter breeze. I see the tree, and from its movement I infer the breeze (I'm inside, so I do not feel it directly). But even more, I see the end product of natural selection. I see branches of a tree with physical characteristics that do more than turn the trees' photosynthesizing needles to the sun and promote passage of nutrients from the trunk and back to it again. I also see a structure neither too rigid nor two easily flexed, a structure adapted to breezes such as this one.

As I watch the branch bob in the breeze, I care little about tomorrow's weather, or even about weather later today (although I shall care about it soon, for I hope to ride my bicycle into town, and it may rain). For now, I think less of these than I do about the weather in my dooryard at this very moment.

But let's leave my dooryard and look to yours. Begin with temperature. Think of that as a measure of heat, rather than a measure of comfort. To simplify, let's talk about a day in springtime, a day when the forecast calls for a high temperature of seventy-one degrees Fahrenheit. Will your dooryard reach seventy-one degrees? It may or may not, and if you have a thermometer, you can check. This is your local equivalent of global temperature. Some parts of your dooryard may never get close to that forecast high and others might surpass it by quite a bit.

It's easy to see why. Begin by looking at anything in direct sunlight—the bark of a tree, soil you've tilled for planting, your bare arm. Each of these may, at any given moment, exceed seventy-one degrees (your skin has a higher starting temperature, since you are warm-blooded). By how much? The answer depends on dozens of factors—the aspect of the surface (whether it is pointed toward or away from the sun, and at what angle), the color of the surface (your skin will absorb more heat if it is black and reflect more if it is pinkish-white), the movement of air (a breeze may prevent the accumulation of heat), and other factors.

When you stand on the ground in sunlight, the surface of your body (if you could measure it everywhere) reveals an almost unimaginably complex gradient of temperatures. And we know that intuitively, don't we? We know, even if we don't think of it in quite this way. We learn from childhood to accommodate this gradient in the way we clothe ourselves and in the many subtle movements and adjustments we make whenever we are outdoors. (As a redhead with very light skin I may possibly be more aware of this than others.) My point here is not to dwell on the obvious but to use your experience, and mine, as a reference point for understanding the complex array of temperature gradients in your dooryard on a sunny day, for understanding the physical environment and the biological in contrasting conditions of sun and shade. If, somehow, every discrete temperature in your dooryard could be colored by dyes, one color tone to each degree of temperature, the result would make a Jackson Pollock pale by comparison. And not only that: anything that absorbs heat from the sun will radiate that heat to cooler air and cooler things close by, causing complex convection currents. Every once in a while, you may actually witness a shimmer in the air or in a shadow, refractions of this serenely energetic tableau.

For the most part, though, all of this is subtle, and a wind gust on the order of a few miles per hour can overwhelm much or most of it. But not all of it. In the Sonoran Desert, where saguaro cactuses grow to heights of more than fifty feet, most of the very young saguaros, the recruits, are unseen. This is because the seeds that germinate and take root do so in the shade of a "nurse plant," such as a small paloverde tree, which provides shade and shelter to the tiny cactus as it grows. Later in life, the cactus will overwhelm the nurse and displace it altogether. But for a time, it needs a certain kind of weather, and the nurse plant provides it. In your dooryard, there are countless pairings of organisms at this very moment. Experienced gardeners understand all of this, and plan their plantings accordingly.

As a way of getting a sense of this weather, this microscale weather, it might be worth spending a day making some measurements. (I call this "microscale weather" to distinguish it from the more familiar notion of microclimate, which is a somewhat different concept.) What is the temperature on the south-facing bark of the oak in your yard, and what's the temperature on the north side? How do the east- and west-facing surfaces differ? What is the ground temperature where there is bare soil? What is it underneath a cover of grass? Perhaps you might invite a local science teacher, or elementary teacher, to bring his class to your dooryard in order to collect dozens of temperature readings over a short period of time (schools may have enough thermometers to provide one per student). On a day when the air is fairly still, what is the air temperature over a piece of open ground at one foot aboveground, two feet, three feet, and so forth? What happens when puffy, cumulous clouds pass overhead? And what happens to all of this at night, after the sun goes down?

You can go beyond temperature to examine other variable weather measures, such as relative humidity or precipitation. Understanding weather at this microscale, the dooryard scale, will enrich your understanding of the phenological symphony in your dooryard. But, as I've conceded, the microscale is easily displaced by something larger.

Also, learn to see weather *trends* that are reflected in the ecology of your dooryard. There is a range of folklore attached to weather, and much of it is related in some way to phenology, whether the relation is empirically verifiable or entirely mythical. A good source of weather lore is the work of the artist Eric Sloane, whose several books on weather are a delight to those of us who learn efficiently when knowledge is presented pictorially.

I remember perusing my father's copies of Eric Sloane, learning weather lore, close to the time that he (my father, that is) presented me with a used copy of *Alone* by Rear Admiral Richard E. Byrd. I was probably about eight years old at the time. I've never known quite why he gave it to me, whether he was sending a message of some sort. Certainly, there were messages in the transfer of such a text, from father to son, realized or not. I read it, absorbed it, and in some ways have endeavored to live it. It is a book about solitary commitment to scientific knowledge and to self-knowledge, a necessary if unintended commentary on *Walden*.

On a dark winter night during the Antarctic winter of 1934, Byrd climbed the ladder from his cabin buried in the Ross Ice Shelf and made his way out the hatch into the long winter night, where he cleared the snow that choked the cups of his anemometer. He admired the aurora australis and went for a walk, taking care to make his way from flag to flag along a walkway he established for this purpose—to get exercise without getting lost in the polar darkness. He called his temporary home Advance Base, and he kept careful, daily records. Alas, carbon monoxide in his cabin would nearly poison him, and he would be rescued from certain death by the men he had left at Little America, miles to the north, who had been under orders not to organize such a rescue attempt, whatever the cause might be.

Today, weather data collected by Byrd and by thousands of people like him are essential for studying changing climate. The data are not phenological, but they are seasonal. Byrd was only one among hundreds of explorers and scientists who risked life and limb to collect weather data and impressions at remote points on the surface of the earth in an effort to understand our planet. Did weather cause malaria? Some thought as much, until the vector proved to be mosquitoes. Would it be possible to make passage across oceans, whether by ship or later by plane, safe from weather calamities? In time, over the early part of the twentieth century, the weather at sea far from shores became knowable and known.

Byrd was part of that growth in what was knowable. An aviator—he was a contender for the prize that Lindbergh won by crossing the Atlantic in *The Spirit of St. Louis*, as well as the leader of crews aboard the first flights over the North *and* South Poles—Byrd was accustomed to using weather data and forecasting as a means for assuring safe flying conditions. Although it predated aviation by several decades, the National Weather Service had begun, by 1926, to establish weather stations at airports across the United States, supplemented by a network of amateur weather observers who collected weather data on farms and in backyards.

But on that dark winter night in 1933, Byrd was making observations about weather where he was temporarily living, his frozen dooryard. He knew it would be cold, but otherwise he had few presuppositions about the links between his weather that day and the weather in the weeks and months to come. This is a good frame of mind in which to make weather observations. You can make the same kinds of observations that Byrd made, and with something like the spirit of adventure, since we do not, in fact, know what novelties the weather in the advancing years of anthropogenic climate change will bring. Your dooryard is probably not as uncomfortably situated as Byrd's Advance Base, but it is very much on the leading edge of discovery.

Much like phenological observations, so too weather observations and record keeping are very much a matter of what you wish to do. The most basic phenological data you collect should be accompanied by simple weather data, such as temperature and conditions (raining, sunny, light fog). But careful phenological observations, such as budbursts or flowering, or first appearances of birds or insect hatches, are made richer by including specifics about weather, not just temperature, but what kind of cloud cover, if any, barometric pressure, wind speed and direction, and precipitation.

You can get much of these data from currently available weather instruments, some of which are designed to record to a computer. Or, if you want to keep things analog, there are instruments that have been around for tens of years and are still serviceable. Other observations can be made with eyes and ears, such as cloud type and percentage of cloud cover.

As in the case of phenological observations, there are opportunities to participate in citizen science as a weather observer. If this interests you, there are standards for calibrating and siting equipment and protocols for reporting. But if you want to ease into making and recording weather observations, you can begin by measuring or observing the aforementioned characteristics of weather (temperature over the course of the day, barometric pressure, wind speed and direction, precipitation, percentage and kind of cloud cover, and relative humidity).

Wind. Whether a breeze or a gust, I tend to think that wind at ground level is about *me*, a narrow stream of air like the focused breath one expels to blow out candles on a birthday cake. Wind seems personal in a way that clouds, snow, rain, and sunshine do not. But I'm wrong when I think this way, sometimes profoundly so. Winds are often movements of large masses of air, sometimes vast in scale, with pockets of local turbulence.

My problem is that I can't see this unless there are clear signals, such as trees responding to variations in wind speeds, or—if I'm sailing—the ripples across the water that show me where there's wind. You see it in the first few minutes of a sailplane flight, when the tow plane rises and falls ahead of you as it encounters a thermal. There are other clues, but they are harder to pick out, and they are counterintuitive. One of these clues is clear air. The Grand Canyon, which is often filled with grubby air (due to coal-fired power-generation plants nearby), has a crystal clarity and wonderful visibility on days with high wind, the kind that can flatten a tent.

But one doesn't see the wind directly, so you need an instrument or two, which can be either simple and intuitive or pricey (but not all that complex). The standard instruments are an indicator and an anemometer with spinning cups or a propeller. These were once connected to switches and lights or counters but now tend to have sensors that send a digital signal to software. They should be placed so as to get an unobstructed "view" of local winds, away from sources of turbulence. But to start, a simple length of yarn (sailors call this a "telltale") will reveal direction and a sense of wind speed. A decorative weathervane also indicates wind direction but not a sense of wind speed. Weathervanes *point toward* the direction that wind is coming *from*, and it is recorded this way. A telltale points the other direction, and the recorded observation must be reversed.

Climate change will bring alterations in the directions and velocities of local winds, the winds you'll experience in your dooryard, but it is difficult to say what those changes will be. In an overly simplistic sense, a sort of hunch, the rise in global temperature should slightly decrease the gradients between air masses on a planetary scale. This should mean winds of lower velocity, even if hurricanes become stronger (with higher wind speeds) or more frequent. But such hunches are only that. Changes in the characteristics of winds—velocities, directions, frequency—are something to be observed so that we learn what to expect.

Winds aloft are another matter, and they move in a couple of ways, one obvious, one not. The obvious movement can be seen whenever there are clouds, passing and changing across the sky and casting shadows that race across the landscape. But there are also vertical movements of rising air, and these show up as cumulus clouds and thunderheads.

Clouds. There should be more thunderheads in the future, because increased temperature should bring with it more evaporation and more precipitation. And not just thunderheads, but clouds of all kinds. How clouds fit into the global climate picture is one of the great unknowns of climate science and is another instance of science and understanding that is still unfolding. Local observations of cloud types, reported through existing weather networks, will help provide greater understanding.

Unlike winds, clouds are observable without the use of instruments and delightful for being so (although the direction in which they are moving requires an instrument, about which I will say something shortly). Meteorologists have devised a classification of cloud types, sorted primarily by where in the vertical layers of the atmosphere they form, accumulate, and move. From the lowest to the highest, the primary classifications are nimbostratus, stratocumulus, stratus, altocumulus, altostratus, cirrocumulus, cirrostratus, and cirrus clouds (fig. 10.2). Will some clouds appear more frequently, while others become rarer? This is what cannot be predicted with confidence. The first order of business is to make observations and, then, to explain them.

But let's sort them out, first. Clouds aren't merely water vapor, the gaseous state that water can take. Water vapor is often transparent. It is also a significant greenhouse gas. It is present in the air we breathe, is measured as relative humidity, and is a necessary precondition for clouds. Clouds form when water vapor, carried aloft by convection, cools and condenses on solid particles of some kind, microscopic bits of dust. Winds then transport the condensed water in clouds (in liquid/droplet form or in solid/ice form) to locations where the water falls back to Earth as precipitation. Because the atmosphere is generally warmest near the surface of the earth and coolest at altitudes, low clouds tend to be wet clouds and high clouds are icy.

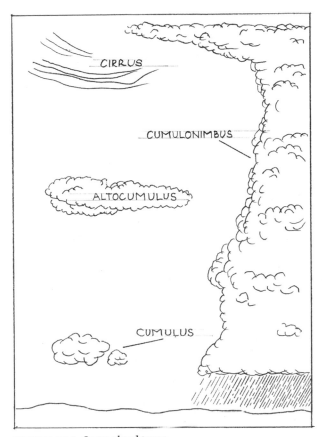

FIGURE 10.2. Some cloud types.

The roles that clouds play in anthropogenic climate change have yet to be sorted out. Climatologists tend to talk about them as a balancing act of positive and negative feedbacks, but let's set that terminology aside and think about various roles they play in the overall energy balance.

First, as soon as a cloud forms in daytime, it begins to absorb some solar radiation, to let some radiation pass through, and to reflect solar radiation back out into space. How much of each of these it does depends on myriad details. At night, when solar energy isn't there, clouds absorb energy radiated from the surface and radiate some of it back to Earth. This is the reason that, all other things being equal, a cloudy night tends to be warmer than a starry night. But some of the absorbed heat is radiated out into space. So the cloud is a kind of waystation for energy before the energy is redirected—back to Earth or out into space. This is where the science is at the edge of discovery. What are clouds doing? What does their activity depend on? What is the overall role of clouds in changing climate?

The National Aeronautics and Space Administration is currently involved in studying clouds by using satellites, some in stationary geosynchronous orbits and some that pass over points on the surface of the earth at regular intervals. Through their SCOOL program (Students' Cloud Observations On-Line), NASA also gathers ground-based observations from school-aged citizen scientists who time their cloud watching to coincide with the passage of a satellite overhead.

But your own observations of clouds will enrich your phenological observations, perhaps someday providing clues to you or to someone else for understanding climatic changes. Sometimes, more than one kind of cloud is visible at a given time, sometimes two or three levels maybe be active with clouds.

In addition to observing and recording cloud types, you may also want to observe their motion, recording the direction they are headed. You can make approximations or make a simple device with a circular mirror, oriented toward the compass directions (N, NNE, NE, ENE, E, for instance). A mirror also makes it simpler to estimate percentage of cloud cover.

Precipitation. Some clouds produce rain. Meteorologists use tipping bucket rain gauges to measure rainfall (in inches), but you can make a simple gauge using a funnel and a collecting tube, calibrated to provide an accurate measurement. Snowfall can be measured simply, with rulers in a few places, and averaged. The Community Collaborative Rain, Hail and Snow Network, is a citizen science network that collects precipitation data.

Temperature. The ideal instrument for recording outdoor temperature is a sensor, placed in ventilated shade about five feet aboveground. This might be a fairly inexpensive indoor/outdoor thermometer, which you can read from inside your home and from which you can record daily highs or lows. Or it can be a more expensive devise that transmits data to a digital recording device, giving you a full picture of temperatures throughout the day or night. Given the importance of temperature for some phenological events, this more thorough treatment of temperature recording may be warranted.

Barometric pressure. There is no better quantitative measure for indicating weather than barometric pressure, and no better predictor of the coming trend. Rising pressure usually means the weather will be fair in due course; falling pressure warns of instability. Members of the Royal Society of London spoke of the weight and spring of air as early as the 1660s and devised elaborate experiments using an air pump (that is, a vacuum pump) to change barometric pressures. In the United States, barometric pressure is reported in inches of mercury, in parallel with degrees Fahrenheit for temperature.

A simple, calibrated barometer is sufficient for reading barometric pressure, but a device that records pressures continuously is more fun.

Relative humidity. There are inexpensive, digital devices for recording relative humidity. But there is nothing like using a sling psychrometer for feeling like a true citizen scientist. A sling psychrometer is a device with two thermometers, a wet bulb and a dry bulb, that is whirled (on a leash) to encourage evaporation on the wet-bulb thermometer. You compare the measured temperatures on the two bulbs to get relative humidity.

Snow cover and snowpack. Snowpack is a major climate consideration in the Western states, and anthropogenic climate change will have an effect on it. All of the water sources in urban California, as well as the highly productive agricultural regions of the Central and Imperial Valleys depend on runoff from snowpack in the Sierra Nevada, the Rocky Mountains, and lesser ranges such as the Wasatch and the Panamints. Snowpack is an important reservoir for storing water for parts of the year, and this fact was one of the justifications for turning over vast areas to the public domain for stewardship by the U.S. Forest Service. Changes in snowpack will have major, not gradual, consequences for many Western cities and for our national food supply.

But in any place where precipitation is held in reserve in frozen form for part of the year, you may see ecological changes in response to climatic changes. The areas and numbers aren't as vast in some places, in the east for instance. But lakes, ponds, rivers, streams, springs, and groundwater may hold reduced volumes of water during several months in some years, relative to what was once normal, transferring that effect to plants and animals. If you live in a place that gets snow every year, or most years, a record of snow cover in your dooryard is worth keeping. Make a note of the date when snow falls, then track the percentage of snow cover as it diminishes to 50 percent and to nothing. If subsequent snow falls, measure the total snowpack about once a week until it is gone. Measure in three places and get an average to record.

Snow cover isn't only a reservoir for water. It is also important for the earth's energy balance, as the high albedo of snow (or ice)—its reflectivity—sends solar energy back into space. When there is diminished snow or ice cover, whether on land or at sea, the earth absorbs more heat. This is a significant positive feedback in the energy budget process. Important as this is, it is a side issue in your dooryard, where the ecological consequences of reduced snowpack is of more concern.

But what about increased snowpacks, such as the "polar vortex" in northeastern states during the winters of 2014 and 2015? It remains to be seen whether these winters, with major snows and thick snowpacks, were anomalous or are an emerging climate feature. Certainly it's the case that both of these winters were warmer and dryer for the whole of the United States. A colder and snowier Northeast in years to come would not be inconsistent with warming on a continental or global scale, but it would certainly be counterintuitive and hard on New Englanders and New Yorkers. And, should *that* be the new normal, people living in this region will want to be alert to phenological changes that lead to ecological change.

Variability and volatility. One aspect of weather that I like to notice, and for which there is no regular category recognized by the National Weather Service, the Weather Channel, or elsewhere, is daily variability or volatility. How much did the weather change, and how often? There are days that include blue skies and thunderstorms; rain, hail, and sunburn; shirtsleeves and snow. Meteorologists do pay attention to variability in weather—and climate—but it is not generally reported, and there is no widely recognized metric for expressing it. For this reason, on a day that includes a wide variety of weather types, there is some value in capturing this in a simple prose description.

Also deserving of descriptions in prose: extreme weather. Why not start with memories of extreme weather to establish a baseline in your experience? I experienced the outbreak of tornadoes in and around Xenia, Ohio, in 1974, for which I was close enough nearby to have felt unsettled but not close enough to really know what it means to experience a tornado. More about that in due course. My other extreme stories are of the blizzard in Boston in 1978 and Hurricane Gloria in 1985. I experienced both of these as larks, really, although the latter is more clearly drawn in my mind. As it passed through my town, with bands of rain alternating with merely moist, warm air, I wanted to be outside experiencing it. But I was home with my then two-year-old son, who wanted nothing to do with a father who attended to the weather rather than to him in such a crisis I kept the doors and windows closed. The power went out and stayed out for five days. We cooked on a wood stove and in the evenings enjoyed lighting from candles and a Coleman lantern, dubbed "the cheery light."

So, my extreme weather stories aren't so extreme. As if that weren't enough, I don't seem to get the full benefit of notorious inclement or even unpleasant weather. I've never been cold in summer in San Francisco, and I basked in the sun at a sidewalk café in London in early autumn.

From the earliest days of concern about anthropogenic climate change, which was usually, in those bygone days, called the greenhouse effect, activists, climate scientists, and the public have showed concern about an increase in extreme weather events—hurricanes, droughts, and very hot days and nights—as likely consequences of increased carbon dioxide in the atmosphere. But in an instance of frustrating rigor, climate scientists have cautioned that no single weather event should be attributed to anthropogenic climate change alone. Activists, like Bill McKibben, eventually lost patience with this caution, arguing that it too easily served the aims of those who seed doubt about climate change.

Climate scientists, for their part, are working on models that may let them say something about the role of anthropogenic climate change in fostering extreme weather events. In the meantime, the attitude among most climate scientists has been to expect extreme weather events in greater frequency and of greater magnitude.

In keeping with the aims of this book, I will talk about extreme weather that you might experience in your dooryard and about how to add such experiences to your understanding of all that will be happening there.

Extreme weather includes hurricanes, droughts, wildfires, extreme heat and extreme cold, extreme snow and ice storms, other storms, and extreme rainfall. The National Academy of Sciences recently ranked these events according to the confidence that climate scientists have in attributing them to human-caused climate changed.[1] From greatest to least confidence, the rankings are as follows:

Extreme cold (decreasing frequency and magnitude)
Extreme heat (increasing frequency and magnitude)
Drought
Extreme snow and ice, extreme rainfall
Hurricanes
Severe convection storms, wildfires

To say that science has low confidence in attributing wildfires to anthropogenic climate change, to take one example, does not mean that wildfires are unexplainable or that they lack a human fingerprint. Wildfires are (1) often caused by humans, (2) exacerbated by human action over decades (primarily through fire suppression), and (3) are made more dangerous and costly by ever-lengthening interfaces between wildlands and our built environment. Climate is but a single factor in all this and difficult to separate from others.

---

1 Committee on Extreme Weather Events and Climate Change Attribution; Board on Atmospheric Sciences and Climate; Division on Earth and Life Studies; National Academies of Sciences, Engineering, and Medicine, *Attribution of Extreme Weather Events in the Context of Climate Change* (Washington, DC: National Academies Press, 2016), fig. S.4.

In events of extreme heat, both the maximum and minimum daily temperatures have increased. Heat waves have become more common now in the United States than they were in the period before 1950 and will become more common still, with higher maximum and minimum temperatures.

One might guess that warming would lead to fewer and less severe occurrences of extreme cold, and the records over the past sixty years bear this out, although "fewer" does not mean "never."

Droughts can be expected to be more common in future years, but attributing any particular drought to climate change is difficult for a long list of reasons. Droughts vary in length and in the seasons in which they occur. (There are, for instance, snow droughts affecting snowpack.) Like wildfires, droughts are sometimes caused by multiple factors. The Dust Bowl might well be called the "great American drought," a record drought that added a second and devastating layer of tragedy for many Americans who were already overcome by the Great Depression. In some ways, it is the best example before anthropogenic climate change of the consequences that human action can have on environment. Individual farmers and policy makers alike learned that climate, combined with settlement and farming practices unsuited to the landscape, can lead to economic displacement and environmental degradation of the highest order. They also learned that warnings of catastrophe are sometimes well-informed.

In the early decades of westward settlement in the United States, pioneers bypassed the Great Plains for fields and pastures farther west. Beginning with the Homestead Act of 1862, the federal government began to promote settlement in the Great Plains. In retrospect, this seems ill-advised, but there were elements within the government that could have prevented disaster. John Wesley Powell, the second director of the U.S. Geological Survey, warned that the Homestead Act was flawed in that it treated all land as though each 160-acre parcel had roughly the same access to water, either from rains or through irrigation. Powell was ignored, in part because the Plains states were unusually well-watered from the late 1870s through the early 1890s.

To gain deed to the 160 acres they claimed, settlers had to live in place for five years and improve the land. This they did by bringing Eastern farming practices, such as deep plowing, to a more arid landscape. The unsuitability of such practice became evident when drought struck the region in the 1930s. Without the presettlement root structures of native plants, and without moisture to substitute for the natural condition, the topsoil simply blew away in windstorms. Most of the homesteaders or their children moved on. But enlightened farm policies helped in the recovery of the landscape.

Comparable consequences of drought in California did not fully materialize recently because of the El Niño of 2015–16, and this underlines the difficulty of attributing drought to any single factor.

One might expect extremes of precipitation to be easily attributed to climate change, and indeed, there is some confidence that heavy downpours and snowstorms (and blizzards) can be associated with anthropogenic climate change, particularly in North America. As for hurricanes, there is a correlation between warming seas and increases in the sizes and number of hurricanes, but there are other factors that limit the number of hurricanes in a given season.

Finally, there are tornadoes, about which climate scientists feel the least confidence in linking a weather event to anthropogenic climate change. The wail of sirens during the Xenia outbreak of April 1974, also known as the Super Outbreak, is a sound I'll never forget. I lived in nearby Cincinnati at the time, and tornadoes touched down near my home there; but the devastation close to me was nothing like what hit Xenia, Ohio, about fifty miles to the northeast.

On April 3 and 4, 148 tornadoes appeared in twelve states plus Canada. In Ohio, tornadoes followed a path from Cincinnati through Xenia, where an F5 tornado, the most powerful in the Fujita scale of zero to five, destroyed a wide swath of structures in the town, injuring more than a thousand people and killing thirty-two. While I heard sirens in Cincinnati, none were heard in Xenia. The town had not installed sirens.

Individual tornadoes are major calamities for those who get caught in their paths. But large outbreaks, such as the Xenia outbreak and the outbreak of April 2011, can also tax regionally shared resources.

Whether a weather event can be clearly attributed to anthropogenic climate chance is of interest to climate scientists and to policy makers but has little bearing on anyone who recognizes that human action in the environment plays a role in climate, through weather events or through phenological change. There is value in taking time to describe all extreme weather events, including the period of time leading up to each event (if forewarned) and the period following it. Phenological observations of all kinds are especially of interest in the aftermath of an event.

PART 3

11

Ground Truth

Welcome to the Anthropocene.

This greeting is belated, of course. We have all been here the whole of our lives, without knowing it for most of our days. The Anthropocene is a span of geological and evolutionary time (technically, an epoch) during which humans have had an outsized role in transforming the natural world, most of that time without knowledge and awareness of the consequences of their actions.

Those who use the term argue about when the epoch begins. Does it start with the Industrial Revolution, late in the eighteenth century, hastening around 1850? Or does it make sense to push the date back in time to the first appearances of agriculture in scattered populations, as much as ten thousand years before the present, or even earlier? It matters more to scientists and scholars than it does to you, whenever you step out into your dooryard. You're in the Anthropocene. Choices made by many generations of people just like yourself, just like me, influence what you come across there. You'll find the influences in small things, mostly. You'll find them in things that are very easy to overlook—a lilac blooming a week earlier than it did a couple of decades ago, for instance. But look a little further afield and you'll find them in far greater ways as well.

Sea levels are rising. In the absence of major changes in the way that we burn fossil fuels, they will rise alarmingly, enough to inundate major population centers—some are cities but others are critical agricultural lands—throughout the world. It is unlikely that rising sea levels will be gentle affairs, permitting time for people to slip away quietly, rebuilding or replanting on higher ground. On the contrary, sea level rise will go hand in hand with storms and storm surges, like Hurricane Sandy

FIGURE 11.1. The view from the site of Henry David Thoreau's doorway in late November 2013 at Walden Pond, Concord, Massachusetts.

in 2012. When sea levels rise, they will do so along coastlines that have been redrawn by extreme weather events and by hastened geological processes like erosion.

Glaciers will surge and drop into the sea. The oceans will grow more acidic. Corals will dissolve in the resultant fizz. And most important, droughts whose total magnitude may dwarf anything yet experienced will send refugees fleeing in various regions around the world. As with so many events, it will be difficult to finger anthropocentric climate change as a singular cause of the mayhem, but it will be a major contributing factor.

In a worst case scenario, grounded ice in Greenland will melt so quickly that the infusion of freshwater into the North Atlantic Ocean will change the density of local seawater, possibly bringing change to global oceanic currents because of the critical flow of seawater in just this place.

I have chosen to keep such projections and predictions out of this book up to now. There are hundreds of books predicting the many scary consequences of climate change. I do not demur from these, and I have no good news to report. I've simply chosen to talk about something else.

Historians know that there are trends in human history that dwarf events we like to think of as "historic." These trends unfold over decades; some bring wholesale change to the course of history, with effects that last for centuries. One such is known as the Columbian Exchange, or European Contact. That's the moment when people from Europe crossed the Atlantic Ocean, bringing diseases (such as smallpox and syphilis) to what would come to be known as the Americas, decimating the indigenous populations here and setting the stage for conquest.

Other events occur over even longer periods of time and have a mixture of positive and negative consequences. One such was the spread of literacy, beginning in the seventeenth century, as a response to the Protestant Reformation, with its emphasis on reading and interpreting the Bible, one sinner at a time.

Then there was the Black Death of the fourteenth century, which killed tens of millions in Europe.

Anthropocentric climate change is an event on this scale, perhaps even greater than any of these by an order of magnitude. For unlike the Black Death, the spread of literacy, and the Columbian Exchange, anthropocentric climate change is reshaping the complete biota, and the physical earth, globally, over a very short period of time. And in addition to a difference in magnitude, this event differs from these others in two profoundly important ways. First, it is reversible and it is within our power to reverse it. Second, because we are aware of it, we can be quite attentive to how it unfolds. We can learn from it.

It is not too late to stop it and reverse it. That can be done, but not once and for all. Thoughtful people across the planet have been searching for decades to find ways to reduce the quantities of carbon that make their way into the atmosphere. Continued commitments from the peoples and governments of the world, including those of the United States, may, in time, prevent the worst among the several projected calamities from occurring. Even then, climate change, and the need to confront both its causes and its consequences, will probably require decades of action, ingenuity, and hope amid setbacks. New technologies may render it simpler to reduce carbon emissions from making their way into the atmosphere. But so long as carbon remains in the ground, especially in the form of coal, there will be a temptation to make use of it. Thus it is unlikely that this is a problem that will be solved once and for all time.

Why do I think this? I take my guarded pessimism, for that's what it is, from a somewhat different realm, that of economics. Economists learned, in the middle decades of the twentieth century, just what was required in order to recover from a global depression. It takes spending by governments, the customers of last resort, on a fairly large scale. In the 1940s, this spending, Keynesian spending as economists call it, was justified by the Second World War. But spending on a similar scale in peacetime would have been appropriate in the 1930s. This was the lesson learned in the 1940s, captured in one of the primary textbooks for economists (by Paul Samuelson) and then forgotten by the time of the downturn that came in 2008. Economists such as Paul Krugman decried the inadequate economic stimulus—it may have been about half what was needed—that was passed by Congress and signed by President Obama in 2009. The result was anemic growth. Europe fared even less well. Thus, a key lesson of economics and policy making was forgotten for a variety of reasons. Surely one of those reasons was avarice, but there were other reasons.

Are we likely to keep our memories intact about climate change? It is certainly my hope, the core of this book, that we can and will. But even with as profound a basis for memory as tens of thousands of phenological records, reducing carbon emissions and keeping them as low as possible will be a problem over the *longue durée*.

The reason for this, the foundation for the problem, is the philosophical ground on which American institutions of governance are built. By this, I'm referring to the notion of God-given and inalienable rights, cited by Thomas Jefferson in the Declaration of Independence but taken by Jefferson from the philosophical work of John Locke's *Two Treatises*. The trouble with Locke is that his thinking came well in advance of ideas about the changing nature of nature itself and of its limits. To provide a major example, Locke knew nothing of the idea of extinction, an idea that would not arrive in the world of ideas until Georges Cuvier established it in 1800. For Locke, nature was a constant, given by God. By the time Cuvier issued his caveat—and he did not think of it quite that way—the major works of classical liberalism were written, and nations—the United States, France—were promising to guarantee certain rights of property and certain liberties, no matter what nature might say about the matter.

This is just where we still remain. Virtually all policy designed to forestall the consequences of climate change threatens, in some way, certain liberties and rights to property that are the bedrock of American political culture (never mind the fact that the consequences, unforestalled, deny whole island nations and people living near sea level of their rights and property). And so, from a certain point of view, it is rational to stand against any policy that limits liberty and confiscates property or diminishes its value. Garrett Hardin understood this quite well and explained it in his article "The Tragedy of the Commons."

One of a number of ecologists and biologists who made their concerns public in the 1960s, Hardin was concerned that unchecked human population growth would outpace global food supplies in just a few decades. Although that did not happen, for reasons that include responses to Hardin's concern, the argument he made has a broader application. There were two parts to what Hardin argued; one of them, having to do with "game theory" in economics, also has applications to ecological theory, but I will put it aside in order to consider the other. This has to do with classical economics: when we have a choice to engage in some kind of exchange, dollars for bread for instance, we make a rational choice. It's never fun to give up dollars, but bread tastes good and keeps us alive. So as long as the price is right, not too many dollars for too little bread, we make the exchange. Economists talk of this decision as one having to do with "utility." If bread has a positive utility of 1, and giving up dollars has a negative utility, one of –1, then this is a fair and rational exchange. Most exchanges work this way, but there are some that do not. Among those that do not are environmental things—things that we share in common, but sometimes exchange. Economists call these "common pool resources," but Hardin compared them to the old notion in New England of a town commons, everyone's dooryard, a plot of grazing land shared by all the townspeople.

As long as townspeople made use of the commons without straining the carrying capacity of the resource—that is, as long as there was more grass to feed cows than there were cows to eat it all—everything was fine. But here, the decision to add a cow to the commons (perhaps a townsperson wanted to produce excess milk so as to make butter or cheese for sale) has a different mathematical model of utility from that of an individual choice. Here, the positive utility of adding a cow might be +1, but the negative utility never approaches –1. The latter, in case of a common pool resource, is shared by all who depend on the resource. If there are thirty people who own cows, then the negative utility is –1/30. What that means is that it is always rational, on purely economic grounds, to add a cow.

And yet we know that it isn't rational on ecological grounds. At some point, there will be more cows than there is grass. The cows will eat all the grass, and then the roots, and the resource will be fully consumed: the tragedy of the commons.

Hardin argued, in his paper, which was published in *Science* in 1968, that in the absence of a market mechanism for preventing catastrophe, rational people needed to regulate certain kinds of economic activity; in particular, activity under conditions of a common pool resource. And that's precisely the case with climate change. Consider the difference in productivity between a worker who has to carry out a task in an air-conditioned room (at, say, sixty-eight degrees Fahrenheit) and another worker who must carry out the same task at an ambient temperature of ninety-five degrees. The cooler worker will carry out the task more economically, at greater speed, will go through more iterations of the task, and will therefore be more productive. If the positive utility of the increased productivity exceeds the negative utility of providing air conditioning (buying an air conditioner and paying for electricity), then it's a rational choice to have a worker who is cool as a cucumber.

The problem here is that one part of this equation hasn't been accounted for. The carbon dioxide produced as a waste product (economists call this an externality) isn't part of the equation. Hardin says: regulate it.

To a climate change activist, the idea that higher productivity is rational is the problem, and it is a problem in perpetuity so long as our political and governmental institutions are grounded in an understanding of nature consistent with Locke, and with the absolute time and space of Isaac Newton, but inconsistent with a conception of natural change, best exemplified by Darwin's *On the Origin of Species*. That is, our political institutions were given form in a time when nature was considered rather static and inexhaustible. We might—or our progeny might—reconstitute our political entities on a new, more Darwinian basis. But until that happens, the battle is to persuade majorities in democratic societies that their interests are best served by giving up a little Newton and accepting a little Darwin.

In the meantime, there will be extinctions caused by climatic changes, just as there have been extinctions caused by loss of habitat, by the outright slaughter of species, and by a variety of other causes. Some of those species might be in your dooryard or pass through it on a fairly regular basis. Your records and those of others who attend to nature may be, in time, all that remains of species that once walked, flew over, or rooted in this planet. Or your records might be among those that may save a species.

But wait. What about emergent technologies? Aren't they our best hope? Energy from the sun? Wind? Tides? Nuclear? These are indeed important, and there is strong hope that continued development will bring the cost of some of them, solar, for instance, below any price that coal may bring.

At present, though, the technologies we are most likely, as citizens and as activists, to confront are certain geo-engineered technologies designed to deal with climate change more or less directly, without demanding changes in patterns of individual energy consumption. These include ideas such as adding sulfides to the atmosphere, or putting reflective materials in orbit around the earth, or collecting carbon with huge vacuum cleaners and placing it in the oceans or some other carbon sink.

So-called climate hackers have had a field day devising a variety of Rube Goldberg–like schemes to cool the planet in an attempt to overcome the consequences of increasing atmospheric carbon dioxide concentrations and associated positive feedbacks, without requiring changes in overall carbon consumption worldwide. In his book *Fixing the Sky*, James R. Fleming, a professor of science, technology, and society at Colby College in Maine, raised important questions about how various climate intervention schemes might (or more likely, might not) work and who might control them. Among the speculative ideas being bandied about are programs to inject sulfate aerosols and other kinds of reflective particles into the upper atmosphere using a kind of sulfate cannon or a hose connected to a high-altitude balloon. Such particles would reflect a small percentage of sunlight back into space before it enters the atmosphere and thus emulate the effects of a volcano, which can cool the climate. Fleming has found that such ideas have a long and checkered history, stretching back almost two hundred years, in some cases, and that they have been the proposed solution to a variety of problems. But who would execute them? Who will pay for them? Who will control the thermostat? And who will clean up the mess if, like that miracle substance asbestos, they create more problems than they solve?

Perhaps the defense against warming will come to mirror America's defense against "all foes foreign and domestic" and a substantial fraction of every tax dollar will go toward geoengineering, just as a fraction goes toward military spending today, with vested interests in place to keep spending constant or growing. But that seems unlikely. Meanwhile, as long as fossil fuels remain in the ground, they will draw some to exploit them, while others forget the reasons why they aren't in broader use.

If you have not already done so, it is worthwhile to examine your own carbon footprint by inventorying all of the energy you use over the course of a year. How do you heat or cool your home, or both? How do you light your home and work spaces? How much hot water do you need, and how hot? What do you require in the way of transportation? There are many possible changes you can make in heating, insulation, and cooling your home. Any changes that reduce carbon emissions are also likely to save money as well, although the time required for energy-saving improvements varies. If you rent, your options may be more limited, but perhaps not as limited as they appear. A little research into improvements and subsidies, along with a willingness to put up with some inconvenience during a retrofit *might* be an encouragement for a landlord to make energy improvement retrofits. In some cases, committed renters might volunteer to pay slightly higher rent in exchange for improvements, especially if you are paying the utilities. One place where this might make sense is an upgrade from older single-pane windows to better insulating double-pane windows.

In making choices about how to reduce your carbon footprint, be alert to anything that also increases your bliss. It's a strange notion. Conventional wisdom often has it that what's good for the environment comes at some cost to human happiness. But that doesn't need to be the case. There is perhaps no better way to increase your happiness than to reduce the amount of time you spend every day commuting to and from work, whether by car or using public transportation, as well as reducing the distances you travel to deliver children to and from activities, do regular shopping and other errands, and travel for entertainment. The closer you live to your workplace, especially, the more you will save in money, time, and carbon. And the happier you will be. Americans, especially younger Americans, are recognizing this and taking action.

But individual, voluntary actions pale in comparison to large-scale collective actions. Think of the ordinary incandescent light bulb, famously the product of Thomas Edison's mind, which provided as much heat as it did light. Must every light source also be a heater, emitting heat energy and producing carbon dioxide as a by-product in many cases? Concerned citizens—not some abstract thing called "government"—believed that they should not, and so we have transitioned from the heater-light to more energy efficient lighting, not by individual voluntary action (although there were volunteers who adopted these technologies early) but through collective legislative actions.

Even so, individual actions undertaken voluntarily are vital to solving the problems of anthropogenic climate change. One such is simply knowing about it, knowing (for instance) that the atmosphere has, as part of its chemical makeup, more the four hundred parts per million of carbon dioxide and that this measure represents a new and recent condition, a threshold crossed. Yes, it is a fairly abstract quantity, but it's also a round number and easy to recall. In a perfectly rational world, four hundred parts per million of carbon dioxide would be all one needs to know, and all rational people would come together, individually and collectively, to take whatever actions are necessary to reduce carbon dioxide in the atmosphere to some number safely lower than four hundred. Need I point out that we do not live in a perfectly rational world?

Because we do not live in such a world, there is a temptation to indulge in active ignorance. The philosopher Karl Popper developed this distinction: ignorance comes in two flavors. There is passive ignorance, wherein we don't know what we don't know. I am ignorant, for instance, of the details of string theory in physics. I know that it exists, but I simply make no effort to learn much about it. I do not read the scientific literature in theoretical physics that talks about it; I do not indulge in the popular science, either. I don't deny the possible interest or utility of string theory. I just don't know anything about it. It doesn't seem very interesting to me. That is passive ignorance.

Active ignorance is quite a different matter. To be actively ignorant requires a degree of effort so as to dwell in an uninformed absence of awareness.

Anthropogenic climate change provides a textbook case in active ignorance, because an entire industry, professional climate change deniers, the "merchants of doubt" in the title of Naomi Oreskes and Erik Conway's book, has been hard at work for several decades manufacturing the many components that make active ignorance of climate change possible. The merchants of doubt have sown confusion by creating the appearance of scientific debate and controversy where there is none. To indulge in active ignorance, all one has to do is to take up that false debate in one's mind and come away convinced that there is too much uncertainty about climate change to act.

Active ignorance is a near-term strategy for getting through life. It's like writing a check, and then another check, and then another without checking your bank balance. It gives you things you want right now. But in the longer term, it rarely works, and it has unpleasant consequences.

The latter hardly needs saying. But beyond knowing about climate change, knowing what four hundred parts per million means, and possibly having a influence on policy, is it worthwhile to pay attention to it? There are reasons to think it is indeed worthwhile. In no particular order, the reasons are as follows:

- Paying attention to changing nature (and taking careful notes) helps scientists and science.
- Paying attention is a way, a rather productive way, to learn about and appreciate nature.
- Paying attention changes the way you interact with the world, on several levels.
- Paying attention to nature is a way of paying attention to yourself, to who you are.

The first of these, paying attention to changing nature (and taking careful notes), is crucial for understanding the effects that changing climate is having for ecological systems. Among those who believe this to be so is the Intergovernmental Panel on Climate Change, the IPCC, which has highlighted, in its most recent report, the role that phenology can play. The citizen phenologist is a key contributor to this. Have you ever seen a documentary or news report showing a scientist clad in parka and mukluks, taking ice cores or pointing to a stream of meltwater rushing from a glacier? A citizen phenologist, quietly listening for the calls of frogs, is an equivalent image, even if it is not as dramatic. Individual observations of phenophases, even if they are of a single species, when aggregated with the reports of other observers, provide the facts that drive theoretical understanding and make modeling and prediction possible.

While researching this book, I read through hundreds of research articles and book chapters that spoke of predictive models and even predictions for the future, in every aspect of the physical earth and its biota. But while noting the general trends of such predictions, and admiring the care and indeed the cleverness of such articles, I did not commit them either to memory or to notes. Predictions, other than as a general trend, are assuredly *not* the subject of this book. Instead, this is a book about discovery, or discoveries, from hundreds to tens of thousands of them, observed, recorded, and averaged over a long period of time.

Phenological observations and reports are generally of single phenomena, one species at a time. And there is a way of seeing your dooryard as no more than a collection of individuals and individual species. But as anyone with an interest in nature knows, those individuals exist and interact ecologically. Plants do not exist apart from, for instance, the pollinators that help in reproduction or herbivores that hinder growth. Earlier, I compared your dooryard to a symphony. Thinking of it that way is pleasant and useful, but from time to time you should think of your symphony as a collection of real-life individuals. Some of them are out with the flu. Some have a cold, and there's danger that the cold will spread, is spreading. Some are aging and will soon retire. A few are looking to get jobs in another orchestra, in a bigger city. The music still sounds good, but the orchestra producing it is in a state of flux. If they perform Beethoven's Fifth Symphony this year, will it sound as good as it did when they played it ten years ago?

If this sounds like I'm stretching a point, be assured that the reality in your dooryard is many times more complex.

Engage in phenological observations of a few organisms on your phenological trail and you have the raw materials for developing a deeper feeling for nature, not just as it is but as it was and has been. A simple analogy: many American homes, in those that have been around for a handful of decades and passed through a few owners, there is likely to be a doorjamb with pen or pencil or even knife marks charting the growth (in height only!) of generations of children—and grandchildren. These are examples of multidecade records that people routinely keep. The first mark in each series was made as a matter of custom and faith. Those that followed revealed growth and were causes for celebration. Here, alas, the analogy breaks down. Your phenological records

may or may not show change and will yield few opportunities to celebrate. But they *are* the antidote to active ignorance. Your attentiveness
and the products of your attentiveness, accumulating over time, secure
a place of local knowledge against active ignorance.

A better phrase to describe the opposite of active ignorance: ground
truth. I've brought out the phrase before, and here it is again. It's another homely bit of terminology, cooked up by NASA some years ago.
But it is also profound, from the point of view of an epistemologist, a
philosopher devoted to the subject of how we can know the difference
between true propositions and false ones. Epistemologists love the idea
that fine ideas can be grounded, that there is a foundation for establishing truths in the world. What better way of establishing the truth
of climatic change than grounding it in the individual observations of
dooryard phenologists, each of them taking a walk a few times each
week—watching, noting, reporting, as nature in their environs goes
through change. Grounded truth. Ground truth.

I began this book close to a year ago at seven thousand feet, on the
southern Colorado Plateau. As I wrote some of the middle chapters, I
had moved to a high and arid region between mountains in Southern
California. And now, as I type the final words, I look out my front door
at a saltwater cove on the coast of Maine. These are not ideal conditions
for keeping rigorous phenological records, and while I've seen plenty of
change in climate, my mobility was a far greater a factor in promoting
them than carbon in the atmosphere has been. Still, I have been attentive to my surroundings, which included the forest of ponderosa pines
in Flagstaff, Arizona, a region that has just had a health shot of El Niño–
influenced precipitation, welcomed there after several years of drought
and, in the previous winter, almost no snow at all.

In Southern California, I would step out into my dooryard and see
the sun rise over the Coyote Mountains, drawing shadows in the valleys
of Mount San Jacinto, the Santa Rosas, Mount Cahuilla, and the long
ridgeline of Mount Palomar before setting on the other side of the Santa
Ana coastal range, including Saddleback—the twin mountain tops of
Modjeska and Santiago. I encountered black widow spiders and taran-

FIGURE 11.2. The view from the author's front door in late October 2015, Doughty Cove, Harpswell, Maine.

tulas, swallows, a rich brew of shore and waterbirds clustered around an artificial lake, and—one morning that I'll not soon forget—a bobcat waiting with the special patience that only cats have, just a few feet from the window behind my desk, hunting rabbits in my dooryard. Even without that particular thrill, the sound of red-winged blackbirds in the reeds surrounding the lake were a daily joy. Had I stayed longer I might have encountered a mountain lion, neighbors there having reported seeing lion tracks in their dooryards.

Now, as the year comes to an end, I see many ducks and cormorants, along with one sighting of a pileated woodpecker. More tellingly, I am aware, day in and day out, of the rise and fall of the tide in the cove in front of my door, another sort of time cycle altogether.

Entirely by accident, my dooryard visually resembles Henry Thoreau's. Were I to take my photo of Walden Pond, snapped from the exact location of Henry's front door, and a photo of my dooryard here on Sebascodegan Island (better known to locals as Great Island), place them in a drawer for a decade or so, and then try to figure out which is

Henry's dooryard and which is mine, I might not be able to distinguish them. But that only works for the photos taken at high tide. At low tide, my view looks out over a brackish mudflat, with a small stream down the center of it. Walden Pond, of course, was not tidal.

The tides meter my life in a diurnal pattern. Even so, I noticed the change from late summer to fall to early winter, but the dominant trees are evergreen. The deciduous trees are mostly oaks, which do not have the brilliant fall colors of maples and other trees on the mainland, not far away. Seasonal change simply wasn't as stark here as the rise and fall, rise and fall of sea level in my little cove.

No matter. The tides make me aware of the precariousness of my perch. In a century's time, perhaps less, my dooryard may be under water at high tide, and the home that I'm renting will be among the tens of thousands of coastal ruins that future kayakers and canoeists may visit, wondering how the builders of such places could be so foolish as to keep pumping carbon into the atmosphere.

Let us not make that so.

Acknowledgments

This book, or more correctly a quick sketch of it, would be little more than a note in my journal were it not for Christie Henry and the University of Chicago Press, and so I thank and acknowledge her, and them, first and deeply. Christie is there no longer; she defected from Chicago to head up a university press at some Ivy League institution, somewhere in the middle of New Jersey. But while at the University of Chicago Press she provided enthusiastic support and direction, and the book would not exist without her unflagging encouragement. I'm relieved that the University of Chicago Press remains in Chicago. Yvonne Zipter edited the manuscript with such care and empathy that her title ought be Editor and Curator.

My former students Katie Zimmerman and Catherine Martini read the earliest draft chapters, making many useful comments and ridding me of any skepticism about the worth of the project. Several anonymous reviewers read the manuscript, in part or in whole, making comments that improved the book immensely. Any errors that remain are mine alone.

Bowdoin College provided the only institutional support I enjoyed while writing the book, when the Department of Environmental Studies there made me a research associate for a year. Over many lunches while at Bowdoin, historian David Hecht kept me from feeling too isolated while I added chapter after chapter to the manuscript. Naomi Oreskes gave inspiration and support while we were both at the University of California, San Diego, as did Christine Hunefeldt at a later stage. Robert S. Westman helped me work through a historiographical issue that was gnawing at me. Most recently, the faculty and students of Lyman Briggs College at Michigan State University have seen to my care and feeding while I taught classes and as the latter stages of publication played out. I'll single out Katie Hinko, Mark Largent, Isaac Record, and Arthur Ward for special thanks. Thomas Summerhill, of Michigan State University's history department, played a similar role.

Hundreds of students at the University of California, San Diego, and dozens at the University of New England spent untold hours keeping phenological journals, none more enthusiastically than Yasmin Shaddox and Alisha Utter. I thank all my students, but especially Yasmin and Alisha, for showing me that keeping a phenological journal was worth the candle.

A historian of science is lost without a peer group. Talking and joking into the wee hours at meetings of the History of Science Society, I can usually be found in the company of the aforementioned Katie Zimmerman plus Vassiliki Betty Smocovitis, Constance Clark, and Dawn Mooney Digrius.

Were it not for the scores of hours that Martin Rudwick has devoted to my education and acculturation as a scholar, even while he graciously accommodated my independent mien, I would hardly have been able to scrape together the intellectual courage to write this book.

Tammy Hensley was always there when I would have been lost without the sorts of boosts, material and psychological, one needs (and sometimes demands) of a friend.

I want to thank my son, Tristan, not just for giving me the granddaughter to whom this book is dedicated and through whom my personal time horizon was stretched by a generation but also for all the seasons of his childhood, the seasons thereafter, and seasons still to come.

Penultimately, I thank Harold Forsythe. I often counsel students that graduate school is worthwhile if for only one blessing. For it is almost certain that one finds one, often more, but at least one person who shares the experience of graduate training and with whom one can talk on the same intellectual stratum for the rest of one's life. For me, that person was and is Harold Forsythe. I met Harold because he lived in the apartment across the sidewalk from mine when we were learning to be historians. We have talked history, politics, life, and our respective children, ever since.

Finally, for her stabilizing presence in my life, as well as for her keen attention to detail, I thank Emma (*Felis catus*).

Further Reading

Phenology: The current "bible" for those who would practice phe-
nology as an aspect of science is Mark D. Schwartz, editor, *Phenology: An Integrative Environmental Science*, 2nd ed. (Dordrecht: Springer, 2013). Like most edited volumes, it is uneven, but to the serious phenologist, whether professional or citizen scientist, it is indispensible. To see a deep engagement in phenological inquiry there is no better book than Richard B. Primack's *Walden Warming: Climate Change Comes to Thoreau's Woods* (Chicago: University of Chicago Press, 2014), in which Primack, a Boston University biologist recovers Henry David Thoreau's notes and endeavors to locate species that still live in the vicinity of Walden Pond. Primack previous published his phenological discoveries widely in scientific journals and continues to do so.

An older book worth the time spent thumbing through is Helmut Leith's *Phenology and Seasonality Modeling* (New York: Springer-Verlag, 1974).

An Internet search on the word "phenology" turns up hundreds of worthwhile web pages, but the leader for learning about phenology in the United States is USA-NPN.

Henry David Thoreau: In addition to the aforementioned *Walden Warming*, there are many books by and about Thoreau. Start by reading *Walden; or, Life in the Woods* or rereading it if you've read it before. There are many editions, including at least one on the Internet. I recommend *Walden: A Fully Annotated Edition*, edited by Jeffrey S. Cramer (New Haven, CT: Yale University Press, 2004). *Walden* is worth having both as a book on paper and in a digital edition. Most libraries have editions of Thoreau's journals in their collections. For a fine sense of the journal at a mature point in Thoreau's life, I recommend *A Year in*

Thoreau's Journal: 1851, with an introduction by H. Daniel Peck (New York: Penguin Classics, 1993).

Of the biographical works on Thoreau, I commend *Henry Thoreau: A Life of the Mind* by Robert J. Richardson Jr. (Berkeley: University of California Press, 1988). Richardson's style (which he also brought to a biography of Ralph Waldo Emerson) is unusual for scholars; the author breaks Thoreau's life into manageable meditations, each factually grounded around clear topics. *Henry David Thoreau: A Life*, by Laura Dassow Walls (Chicago: University of Chicago Press, 2017), was published too late to have shaped my understanding of the Concord phenologist, but it is likely to be the definitive biography for years to come.

Although written as a work of literary criticism, *Thoreau's Morning Work: Memory and Perception in "A Week on the Concord and Merrimack Rivers," the "Journal," and "Walden"* by H. Daniel Peck (New Haven, CT: Yale University Press, 1990) provides a thoughtful if dense reading of Thoreau's greater purpose in thinking about seasonality and attempting to devise a "kalendar."

Aldo Leopold: *A Sand County Almanac* is must reading, not once but regularly. If your library has a copy of the DVD be sure to watch *Green Fire: Aldo Leopold and a Land Ethic for Our Time*, a documentary that features Leopold's biographer Curt Meine, who wrote the indispensable *Aldo Leopold: His Life and Work* (Madison: University of Wisconsin Press, 1988).

The ecologies of places and how they change: Do not miss out on the books of Bernd Heinrich. *The Trees in My Forest* (New York: Cliff Street Books, 1997) is a good place to start. Also worthwhile is *Second Nature: A Gardener's Education* by Michael Pollan (New York: Atlantic Monthly Press, 1991). The masterwork on this topic, *Changes in the Land: Indians, Colonists, and the Ecology of New England*, was written by the environmental historian William Cronon. It is the apotheosis of clear, scholarly writing. (Try just reading the first sentence of each paragraph. Cronon knows how to write topic sentences and make them count.)

Although it is quite dated, and therefore not to be trusted to be as complete as more recent scholarship, Wallace Stegner's *Beyond the Hundredth Meridian: John Wesley Powell and the Second Opening of the West* (New York: Penguin Books, 1992) makes up with the quality of Stegner's writing and vision what it lacks in currency.

For **phenology networks,** look to the Internet. Here are current web addresses for several leading networks:

> The USA-National Phenology Network (USA-NPN) is the leading network and source of up-to-date information about phenology. Its website lists collaborating networks. www.usanpn.org
>
> Nature's Notebook, part of USA-NPN, is a welcoming portal for beginners. www.naturesnotebook.org
>
> Project Budburst, the network of observers reporting plant phenology posts data on its website. www.budburst.org
>
> FrogWatch USA is a program of the Association of Zoos and Aquariums that collects phenological data on frogs and toads. Its season runs from February through August. Observers are taught to identify frogs and toads by their calls, and report their observations online. www.aza.org/frogwatch
>
> Up-to-date information about the Christmas bird count can be found at www.audubon.org/conservation/science/christmas-bird-count

Clones and other long-lived organisms: Orthogonal to my book but of great interest (and because I mentioned it), be sure to have a look at *The Oldest Living Things in the World* by Rachel Sussman, with essays by Hans Ulrich Obrist and Carl Zimmer (Chicago: University of Chicago Press, 2014).

Climate change science: There are many books that explain the science of climate change. I prefer a textbook that requires no advance knowledge of climate science but is not written for idiots. It is *The Climate Crisis: An Introductory Guide to Climate Change* by David Archer and Stefan Rahmstorf (Cambridge: Cambridge University Press, 2010)

Observation and note taking: Several books by Clare Walker Leslie, including her original *Nature Drawing: A Tool for Learning* (Englewood Cliffs, NJ: Prentice-Hall, 1980), provide guidance not simply for making drawings of plants and animals but for seeing them as well. A more advanced treatment is *Field Notes on Science and Nature*, edited by Michael R. Canfield (Cambridge, MA: Harvard University Press, 2011), in which a dozen scientists discuss not only their own styles of note taking and drawing but also the work of others that influenced them. For a discussion of the importance of observation in science (and an indictment of its diminished role), read *Observation and Ecology: Broadening the Scope of Science to Understand a Complex World* by Rafe Sagarin and Aníbal Pauchard (Washington, DC: Island Press, 2012). Picture Post (picturepost.unh.edu) is a website devoted to repeat photography.

Plants and botany: The book I've preferred for many years, which is both out of print and likely out of date, is *Humanistic Botany* by Oswald Tippo and William Louis Stern (New York: W. W. Norton, 1977). For a careful overview of ethnobotany in a small geographical region, do have a look at two books by Gary Paul Nabhan. The first is *Gathering the Desert*, with illustrations by Paul Mirocha (Tucson: University of Arizona Press, 1985). The other is *The Desert Smells Like Rain: A Naturalist in O'Odham Country* (San Francisco: North Point Press, 1982). Also noteworthy are the several books by Ann Haymond Zwinger, including *The Mysterious Lands: A Naturalist Explores the Four Great Deserts of the Southwest* (Tucson: University of Arizona Press, 1996). I favor Nabhan and Zwinger because the Sonoran Desert was the first place I really began to look at plants. But they are part of a rich, deep, and ongoing tradition of natural history writing focused on plants. Consider, for instance, *Gathering Moss: A Natural and Cultural History of Mosses* by Robin Wall Kimmerer (Corvallis: Oregon State University Press, 2003). There are doubtless similar writers who speak of your region of the country (including, in this context, the aforementioned *Walden Warming* by Richard B. Primack). *Botany Primer* (by P. Guertin, L. Barnett, E. G. Denny, and S. N. Schaffer [2015]) is available as a free PDF file from the USA-NPN at https://www.usanpn.org/files/shared/files/USA -NPN_Botany-Primer.pdf.

Birds: The literature of birds and bird watching is vast. I strongly recommend two books by Scott Weidensaul, *Living on the Wind: Across the Hemisphere with Migratory Birds* (New York: North Point Press, 1999) and *The Ghost with Trembling Wings: Science, Wishful Thinking and the Search for Lost Species* (New York: North Point Press, 2003). A book that deeply influenced my interest in birds and habitat is Peter Matthiessen's *The Wind Birds* (New York: Viking Press, 1973).

Field guides and checklists: There are several series of field guides, each with its own characteristics. The Peterson Guides (such as Roger Tory Peterson's own *Peterson Guide to the Birds of North America*) are the most mature guides still in print, even as they are revised, and emphasize generalized paintings and drawings of species. Sibley's guides (for instance, *The Sibley Guide to Birds of Eastern North America*, 2nd ed.) are quite popular. At the opposite end of the spectrum are the National Audubon Society guides, which use photographs instead of drawings and paintings. If you don't already own guides that you find helpful, I recommend borrowing several guides from the library or from friends and discovering which of them you find most helpful in identifying plants and wildlife. This may vary—you might find that photographs of birds suit you, but that more generalized drawings of wildflowers provide greater clarity in your pursuits. It's possible that you'd like to have two guides, one with drawings and one with photographs. In addition to the Peterson series and the National Audubon series, there are also series by the National Geographic Society (photographs) and the Sierra Club (which are organized by region). More regional guides and checklists are available. A good source for recommendations are local chapters of the Audubon Society (such as, in my state, the Maine Audubon Society, which has a bookstore at its state headquarters).

Weather: For histories of meteorology, see the works of James Rodger Fleming, including his *Meteorology in America, 1800-1870* (Baltimore, MD: Johns Hopkins University Press, 1990). Folksy treatments of weather phenomena and weather lore are found in several books by Eric Sloane, including *Eric Sloane's Weather Book* (Mineola, NY: Dover Publications, 2005) and *Folklore of American Weather* (New York: Hawthorne Books, 1963).

CoCoRaHS, the Community Collaborative Rain, Hail and Snow Network, collects precipitation data from citizen scientists: http://co corahs.org.

SCOOL, the Students' Cloud Observations On-line, is the NASA program that coordinates ground observations of clouds with satellite monitoring. SCOOL is at: https://scool.larc.nasa.gov.

Climate change policy: Although it was written decades ago, Bill McKibben's *The End of Nature* (New York: Random House Trade Paperbacks, 2006) is still the clarion call with respect to understanding climate change and our changing place in nature. *Merchants of Doubt: How a Handful of Scientists Obscured the Truth on Issues from Tobacco Smoke to Global Warming* by Harvard historian Naomi Oreskes and Caltech historian Erik Conway (New York: Bloomsbury Press, 2010) is the essential book for understanding why we are so at sea with respect to fixing the problem of anthropogenic climate change. If you don't read the book, at least watch the documentary that was made from it with its authors' aid and approval.

Index